DRUG TRANSPORT ACROSS THE BLOOD-BRAIN BARRIER

DRUG TRANSPORT ACROSS THE BLOOD-BRAIN BARRIER
IN VITRO AND *IN VIVO* TECHNIQUES

Edited by

A.(Bert) G. de Boer

Leiden/Amsterdam Center for Drug Research,
University of Leiden, The Netherlands

and

W. Sutanto

Leiden/Amsterdam Center for Drug Research,
University of Leiden, The Netherlands

harwood academic publishers
Australia • Canada • China • France • Germany • India • Japan
Luxembourg • Malaysia • The Netherlands • Russia • Singapore
Switzerland • Thailand • United Kingdom

Copyright © 1997 OPA (Overseas Publishers Association) Amsterdam B.V. Published in The Netherlands by Harwood Academic Publishers.

All rights reserved.

No part of this book may be reproduced or utilized in any form or by any means, electronic or mechanical, including photocopying and recording, or by any information storage or retrieval system, without permission in writing from the publisher. Printed in Singapore.

Amsteldijk 166
1st Floor
1079 LH Amsterdam
The Netherlands

British Library Cataloguing in Publication Data

Drug transport across the blood-brain barrier : in vitro
    and in vivo techniques
    1. Brain chemistry   2. Blood-brain barrier   3. Brain
    microdialysis
    I. Boer, A. G. de (Albertus G.)   II. Sutanto, Win
    612.8′22

    ISBN 9057020327

ESKIND BIOMEDICAL LIBRARY

JUN 0 2 1998

VANDERBILT UNIVERSITY
NASHVILLE, TN 37232-8340

CONTENTS

Preface   ix

Contributors   xi

Part A *In Vitro:* **Cell Culture**

I. Introduction   3
    A.G. de Boer and *W. Sutanto*

II. *In vitro* endothelial cell culture: Rat
   II.1. Preparation of primary rat brain endothelial cell culture.
      Modified method of CCW Hughes   5
      N.J. Abbott, P.A. Revest, J. Greenwood, I.A. Romero, M. Nobles, R.J. Rist,
      Z.-D. Reeve-Chen and *M.W.K. Chan*
   II.2. Possibilities, limitations and isolation procedures of rat brain
      microvessel endothelial cell (RMEC) culture techniques   17
      A.G. de Boer, H.E. de Vries, P.J. Gaillard and *D.D. Breimer*
   II.3. Rat brain microvessel endothelial cells: Primary cultures and
      immortalized differentiated cell line   27
      D. Lechardeur, B. Schwartz and *D. Scherman*
   II.4. Drug metabolizing enzyme activities in an *in vitro* model
      of rat blood-brain barrier   37
      A. Minn, D. Gradinaru, G.F. Suleman, M. Chat, C. Bayol-Denizot and
      P. Lagrange
   II.5. Immunohistochemical and electronmicroscopy detections   49
      M.A. Deli, A.C. Szabó, N.T.K. Dung and *F. Joó*

III. *In vitro* endothelial cell culture: Bovine
   III.1. Isolation and primary cultures of bovine brain microvessel
      endothelial cells (BMEC)   59
      A.G. de Boer, P.J. Gaillard, H.E. de Vries and *D.D. Breimer*
   III.2. A co-culture of brain capillary endothelial cells and astrocytes:
      An *in vitro* blood-brain barrier for studying drug transport
      to the brain   69
      B. Dehouck, R. Cecchelli and *M.-P. Dehouck*
   III.3. Primary cultures of bovine brain microvessel endothelial cells   81
      D. Lechardeur, B. Schwartz and *D. Scherman*

IV. *In vitro* endothelial cell culture: Porcine
   IV.1. Preparation of primary culture from newborn pigs   85
      M.A. Deli, C.S. Ábrahám, N.T.K. Dung and *F. Joó*

IV.2. Preparation of endothelial cells in primary cultures obtained
from the brains of 6-month-old pigs 91
*B. Tewes, H. Franke, S. Hellwig, D. Hoheisel, S. Decker, D. Griesche,
T. Tilling, J. Wegener and H.-J. Galla*

V. *In vitro* endothelial cell culture: Human
V.1. Isolation and characterization of human brain endothelial cells 99
M.S.F. Clarke, D.C. West, P. Dias, S. Kumar and P. Kumar
V.2. Methods of isolation and culture of human brain microvessel
endothelium 109
M. Vastag and Z. Nagy

Part B *In Vivo:* **Microdialysis**

I. Introduction 117
A.G. de Boer and W. Sutanto

II. Brain microdialysis: Rat
II.1. Microdialysis as a tool in pharmacokinetic and pharmacodynamic
studies 119
M.R. Bouw, J.S. Sidhu and M. Hammarlund-Udenaes
II.2. Microdialysis measurements of free drug concentrations in
blood and brain 137
C. van Amsterdam, P. Misslin and M. Lemaire
II.3. Transcerebral microdialysis of morphine and morphine
6-glucuronide 149
P. Sandouk, M. Barjavel, F. Stain and J.-M. Scherrmann
II.4. Effect of perfusate tonicity and temperature on microdialysate
AUC values of acetaminophen and atenolol obtained from rat
cortical brain 157
E.C.M. de Lange, M. Danhof, A.G. de Boer and D.D. Breimer

III. Brain microdialysis: Mouse
III.1. Brain microdialysis in the mouse: Determination of biogenic amine
metabolites in the dorsal hippocampus and the nucleus accumbens 165
N. Launay, G. Boschi, R. Rips and J.-M. Scherrmann

IV. Brain microdialysis: Rabbit
IV.1. Microdialysis of subcortical structures in conscious chronic rabbits 173
W.Z. Traczyk, M. Orlowska-Majdak and A. Walczewska

Part C *In Vivo:* **Voltammetry and Microcapsules**

I. Introduction 187
E.C.M. de Lange and A.G. de Boer

II. Uric acid passage across blood-brain barrier to the cerebral cortex
     and corpus striatum of the rat as monitored with differential pulse
     voltammetry (DPV) and fast scan cyclic voltammetry (FCV)   189
     J. Pavlásek, M. Haburčák and *C. Mašaňová*

III. Liposome microcapsules: An experimental model for drug transport
     across the blood-brain barrier (BBB)   201
     F. Mixich and *S. Mihailescu*

Part D Perspectives   215
     A.G. de Boer and *W. Sutanto*

Index   217

PREFACE

The availability of various *in vitro* and *in vivo* techniques has considerably advanced the research on drug transport and metabolism across the blood-brain barrier (BBB). These specialized and sophisticated experimental strategies are of fundamental importance if one is to gain a greater understanding of enhanced and selective drug delivery to the brain.

The reader will find in this book methods for *in vitro* endothelial/astrocyte cell culture models, and for *in vivo* intracerebral microdialysis to study drug transport across the BBB. This book, however, is not merely a laboratory manual consisting of recipes for BBB research; it permits the presentation of the different methods in fine detail, revealing tricks and short cuts that frequently do not appear in the literature. The researcher is well aware that differences (subtle or otherwise) in experimental steps used in different laboratories may influence the outcome of any particular procedure. The book also illustrates the accessibility and the application of the different methods in different species. Background information of the protocol is given in every chapter, which also contains a literature list that the reader may wish to refer to for further information.

The realization of this book would not have been possible without financial support from the Commission of the European Communities. The Editors thank the authors for their contributions, and hope that this volume will be invaluable to basic researchers as well as to those involved in the search for agents suitable for pharmaceutic intervention in the central nervous system.

CONTRIBUTORS

N.J. Abbott
Physiology Group, Biomedical
  Sciences Division
King's College
The Strand
London WC2R 2LS
UK

C.S. Ábrahám
Department of Paediatrics
Albert Szent-Györgi Medical University
P.O. Box 471
6701 Szeged
Hungary

C. van Amsterdam
Biopharmaceutical Department
  507/801, Sandoz Pharma Ltd.
4002 Basel
Switzerland

M. Barjavel
INSERM Unité 26
Hôpital Fernand Widal 200
Rue du Faubourg Saint-Denis
75475 Paris
France

C. Bayol-Denizot
Centre du Médicament
URA CNRS 597
30 rue Linnois
54000 Nancy
France

A.G. de Boer
Division of Pharmacology
Leiden/Amsterdam Center for
  Drug Research
University of Leiden
P.O. Box 9503
2300 RA Leiden
The Netherlands

G. Boschi
INSERM Unité 26
Hôpital Fernand Widal 200
Rue du Faubourg Saint-Denis
75475 Paris
France

M.R. Bouw
Department of Pharmacy
Division of Biopharmaceutics and
  Pharmacokinetics
University of Uppsala
Box 580
751 23 Uppsala
Sweden

D.D. Breimer
Division of Pharmacology
Leiden/Amsterdam Center for
  Drug Research
University of Leiden
P.O. Box 9503
2300 RA Leiden
The Netherlands

R. Cecchelli
Unité 325 INSERM
Institut Pasteur
1 rue Calmette, 59019 Lille
France

M.W.K. Chan
Physiology Group, Biomedical
  Sciences Division
King's College, The Strand
London WC2R 2LS
UK

M. Chat
Centre du Médicament
URA CNRS 597
30 rue Linnois
54000 Nancy
France

M.S.F. Clarke
The Paterson Institute
Christie Hospital
Head Tumour Biology Laboratory
Wilmslow Road
Withington
Manchester M20 9BX
UK

M. Danhof
Division of Pharmacology
Leiden/Amsterdam Center for
  Drug Research
University of Leiden
P.O. Box 9503
2300 RA Leiden
The Netherlands

S. Decker
Institut für Biochemie
Westfälische Wilhelms-Universität
Wilhelm-Klemm-Strasse 2
48149 Münster
Germany

B. Dehouck
Unité 325 INSERM
Institut Pasteur
1 rue Calmette
59019 Lille
France

M.-P. Dehouck
Unité 325 INSERM
Institut Pasteur
1 rue Calmette
59019 Lille
France

M.A. Deli
Laboratory of Molecular Neurobiology
Institute of Biophysics
Biological Research Center of the
  Hungarian Academy of Sciences
P.O. Box 521
6701 Szeged
Hungary

P. Dias
The Paterson Institute
Christie Hospital
Head Tumour Biology Laboratory
Wilmslow Road
Withington
Manchester M20 9BX
UK

N.T.K. Dung
Laboratory of Molecular Neurobiology
Institute of Biophysics
Biological Research Center of the
  Hungarian Academy of Sciences
P.O. Box 521
6701 Szeged
Hungary

H. Franke
Institut für Biochemie
Westfälische Wilhelms-Universität
Wilhelm-Klemm-Strasse 2
48149 Münster
Germany

P.J. Gaillard
Division of Pharmacology
Leiden/Amsterdam Center for
  Drug Research
University of Leiden
P.O. Box 9503
2300 RA Leiden
The Netherlands

H.-J. Galla
Institut für Biochemie
Westfälische Wilhelms-Universität
Wilhelm-Klemm-Strasse 2
48149 Münster
Germany

D. Gradinaru
National Institute of Gerontology and
  Geriatrics "Ana Aslan"
9, Str Manastirea Caldarusani
78178 Bucharest
Romania

J. Greenwood
Department of Clinical Science
Institute of Ophthalmology
11–43 Bath Street
London EC1V 9EL
UK

D. Griesche
Institut für Biochemie
Westfälische Wilhelms-Universität
Wilhelm-Klemm-Strasse 2
48149 Münster
Germany

M. Haburčák
Department of Neurophysiology
Institute of Normal and Pathological
    Physiology
Slovak Academy of Sciences
Sienkiewiczova 1
81371 Bratislava
Slovak Republic

M. Hammarlund-Udenaes
Department of Pharmacy
Division of Biopharmaceutics and
    Pharmacokinetics
University of Uppsala
Box 580
751 23 Uppsala
Sweden

S. Hellwig
Institut für Biochemie
Westfälische Wilhelms-Universität
Wilhelm-Klemm-Strasse 2
48149 Münster
Germany

D. Hoheisel
Institut für Biochemie
Westfälische Wilhelms-Universität
Wilhelm-Klemm-Strasse 2
48149 Münster
Germany

F. Joó
Laboratory of Molecular Neurobiology
Institute of Biophysics
Biological Research Center of the
    Hungarian Academy of Sciences
P.O. Box 521
6701 Szeged
Hungary

P. Kumar
The Manchester Metropolitan University
Faculty of Science and Engineering
Department of Biological Science
Chester Street
Manchester M1 5GD
UK

S. Kumar
Department of Pathology
Medical School
The University
Manchester M13 9PT
UK

P. Lagrange
Centre du Médicament
URA CNRS 597
30 rue Linnois
54000 Nancy
France

E.C.M. de Lange
Division of Pharmacology
Leiden/Amsterdam Center for
    Drug Research
University of Leiden
P.O. Box 9503
2300 RA Leiden
The Netherlands

N. Lauday
INSERM Unité 26
Hôpital Fernand Widal 200
Rue du Faubourg Saint-Denis
75475 Paris
France

D. Lechardeur
UMR 133 CNRS/Rhône-Poulenc
  Rorer
Centre de Recherche de
  Vitry-Alfortville
13 quai Jules Guesde
B.P. 14
94403 Vitry/Seine
France

M. Lemaire
Biopharmaceutical Department
  507/801
Sandoz Pharma Ltd.
4002 Basel
Switzerland

C. Mašánová
Department of Neurophysiology
Institute of Normal and Pathological
  Physiology
Slovak Academy of Sciences
Sienkiewiczova 1
81371 Bratislava
Slovak Republic

S. Mihailescu
Department of Molecular Cell Biology
Faculty of Medicine
4 Petru Rares st.
Craiova 1100
Romania

A. Minn
Centre du Médicament
URA CNRS 597
30 rue Linnois
54000 Nancy
France

P. Misslin
Biopharmaceutical Department
  507/801
Sandoz Pharma Ltd.
4002 Basel
Switzerland

F. Mixich
Department of Molecular Cell Biology
Faculty of Medicine
4 Petru Rares st.
Craiova 1100
Romania

Z. Nagy
Semmelweis Medical University
Stroke Center Budapest
Huvosvolgyi ut 116
1021 Budapest
Hungary

M. Nobles
Department of Pharmacology
University College
Gower Street
London WC1E 6BT
UK

M. Orlowska-Majdak
Department of Physiology
Institute of Physiology and
  Biochemistry
Medical University of Lodz
Ul. Lindleya 3
90-131 Lodz
Poland

J. Pavlásek
Department of Neurophysiology
Institute of Normal and Pathological
  Physiology
Slovak Academy of Sciences
Sienkiewiczova 1
81371 Bratislava
Slovak Republic

Z.-D. Reeve-Chen
Physiology Group, Biomedical
  Sciences Division
King's College
The Strand
London WC2R 2LS
UK

P.A. Revest
Physiology Department
Queen Mary and Westfield College
Mile End Road
London E1 4NS
UK

R. Rips
INSERM Unité 26
Hôpital Fernand Widal 200
Rue du Faubourg Saint-Denis
75475 Paris
France

R.J. Rist
Physiology Group, Biomedical
    Sciences Division
King's College
The Strand
London WC2R 2LS
UK

I.A. Romero
Physiology Group, Biomedical
    Sciences Division
King's College
The Strand
London WC2R 2LS
UK

P. Sandouk
INSERM Unité 26
Hôpital Fernand Widal 200
Rue du Faubourg Saint-Denis
75475 Paris
France

D. Scherman
UMR 133 CNRS/Rhône-Poulenc
    Rorer
Centre de Recherche de
    Vitry-Alfortville
13 quai Jules Guesde
B.P. 14
94403 Vitry/Seine
France

J.-M. Scherrmann
INSERM Unité 26
Hôpital Fernand Widal 200
Rue du Faubourg Saint-Denis
75475 Paris
France

B. Schwartz
UMR 133 CNRS/Rhône-Poulenc Rorer
Centre de Recherche de
    Vitry-Alfortville
13 quai Jules Guesde
B.P. 14
94403 Vitry/Seine
France

J.S. Sidhu
Department of Pharmacy
Division of Biopharmaceutics and
    Pharmacokinetics
University of Uppsala
Box 580
751 23 Uppsala
Sweden

F. Stain
INSERM Unité 26
Hôpital Fernand Widal 200
Rue du Faubourg Saint-Denis
75475 Paris
France

G.F. Suleman
Centre du Médicament
URA CNRS 597
30 rue Linnois
54000 Nancy
France

W. Sutanto
Center for Bio-pharmaceutical
    Sciences
University of Leiden
P.O. Box 9503
2300 RA Leiden
The Netherlands

A.C. Szabó
Laboratory of Molecular Neurobiology
Institute of Biophysics
Biological Research Center of the
  Hungarian Academy of Sciences
P.O. Box 521
6701 Szeged
Hungary

B. Tewes
Institut für Biochemie
Westfälische Wilhelms-Universität
Wilhelm-Klemm-Strasse 2
48149 Münster
Germany

T. Tilling
Institut für Biochemie
Westfälische Wilhelms-Universität
Wilhelm-Klemm-Strasse 2
48149 Münster
Germany

W.Z. Traczyk
Department of Physiology
Institute of Physiology and
  Biochemistry
Medical University of Lodz
Ul. Lindleya 3
90-131 Lodz
Poland

M. Vastag
Semmelweis Medical University
Stroke Center Budapest
Huvosvolgyi ut 116
1021 Budapest
Hungary

H.E. de Vries
Division of Pharmacology
Leiden/Amsterdam Center for
  Drug Research
P.O. Box 9503
2300 RA Leiden
The Netherlands

A. Walczewska
Department of Physiology
Institute of Physiology and
  Biochemistry
Medical University of Lodz
Ul. Lindleya 3
90-131 Lodz
Poland

J. Wegener
Institut für Biochemie
Westfälische Wilhelms-Universität
Wilhelm-Klemm-Strasse 2
48149 Münster
Germany

D.C. West
The Paterson Institute
Christie Hospital
Wilmslow Road
Withington
Manchester M20 9BX
UK

PART A
IN VITRO: CELL CULTURE

I. INTRODUCTION

A.G. de BOER[1] AND W. SUTANTO[1,2]

*Divisions of Pharmacology[1] and Medical Pharmacology[2], Leiden/Amsterdam
Center for Drug Research (LACDR), University of Leiden,
P.O. Box 9503, 2300 RA Leiden, the Netherlands*

The development of procedures to isolate and culture brain microvessel endothelial cells has considerably enhanced the understanding of the transport of drugs across the blood-brain barrier (BBB) and the functionality of the BBB. The development of these systems has evolved in several aspects: on the one hand the optimization of the isolation and culture of cells, and on the other hand the improvement in the materials for culturing the cells. This includes a better knowledge about the culture media, an essential component in the procedure for BBB endothelial cell culture. This all has led to the routine culture of relatively pure confluent high resistance monolayers of BBB endothelial cells. Presently some well-characterized systems are being used to study the BBB *in vitro*. However, several aspect may need to be improved, e.g. the co-culture with pericytes and astrocytes and neurones in order to get a cellular system comparable to the *in vivo* BBB. Nevertheless, the present systems provide reliable tools to study and to predict BBB drug transport and BBB functionality particularly in disease state.

In the first section of this book several isolation procedures and culture systems for BBB endothelial cells are described. These procedures have been optimised, validated and are routinely used in various laboratories involved in a European Community-funded Biomedical Concerted Action programme entitled *Drug Transport across the Blood-Brain Barrier: New Experimental Strategies*.

II.1. PREPARATION OF PRIMARY RAT BRAIN ENDOTHELIAL CELL CULTURE
Modified Method of CCW Hughes

N.J. ABBOTT,[1] P.A. REVEST,[2] J. GREENWOOD,[3] I.A. ROMERO,[1]
M. NOBLES,[4] R.J. RIST,[1] Z.-D. REEVE-CHEN[1] AND M.W.K. CHAN[1]

[1]*Physiology Group, Biomedical Sciences Division, King's College,*
The Strand, London WC2R 2LS, UK
[2]*Physiology Department, Queen Mary & Westfield College,*
Mile End Rd, London E1 4NS, UK
[3]*Department of Clinical Science, Institute of Ophthalmology,*
11–43 Bath St, London EC1V 9EL, UK
[4]*Department of Pharmacology, University College,*
Gower St, London WC1E 6BT, UK

A method for preparing primary cultured rat brain endothelial cells is described. The method is based on protocols developed by CCW Hughes, building on the pioneering work of Bowman *et al.* The method uses grey matter from young Lewis rats, and involves careful removal of meninges and white matter, mechanical dissociation, and a two-stage enzymatic digestion, first to remove non-vascular elements, second to remove pericytes from the vessel fragments. Finally, a Percoll gradient is used to isolate the cleaned microvessel fragments. Endothelial cells grow out from the plated vessel fragments as small colonies, which meet to form a uniform monolayer. Any remaining contaminating cells can be removed by treatment with calcium- and magnesium-free medium, or by killing with Thy 1.1 antibody plus complement. The method yields cultures that are ~95% pure endothelial, and which preserve important blood-brain barrier markers including the transferrin receptor and p-glycoprotein.

INTRODUCTION

Rationale for Developing a Rat Brain Endothelial Cell Culture

There are several reasons for wishing to develop a rat brain endothelial culture.

a. The bulk of our knowledge about the physiology of the blood-brain barrier comes from studies on the rat, so rat cultures offer an excellent opportunity for *in vivo/in vitro* comparisons as a means of elucidating underlying mechanisms.

b. It is increasingly clear that many of the important 'barrier phenotype' characteristics of the brain endothelium are induced by non-endothelial cells, chiefly astrocytes and neurones, but perhaps including pericytes and other perivascular cells; it is useful to have an experimental model in which co-culture experiments can be done using cells from the same species, and in the rat it is relatively easy to prepare other types of cultured cells for co-culture.

c. For characterisation of a primary culture, including cell types for potential co-culture, it is necessary to check purity, expression of characteristic markers etc.. Several convenient antibodies are available for the rat.

d. For some studies of pathology, metabolic disorders and toxicology, it is useful to be able to pre-treat the animal in a certain way, then examine the endothelial cells isolated from the brain after the treatment. Rats are useful models for this kind of study (e.g. effects of diabetes, altered blood glucose, chronic exposure to toxins, genetic manipulation).

e. Rats and mice are extensively used in immunological studies; they are sufficiently similar that preparative methods developed for the rat should be relatively simple to apply to the mouse.

History of Rat Brain Endothelial Preparation

The method is based on that developed by Bowman *et al.* (1981, 1983), and further perfected by C.C.W. Hughes, who spent three years refining the method, trying variants, and characterising the culture that resulted (Hughes & Lantos, 1986, 1989). Chris Hughes helped us set up the method at King's College in 1989, and we have developed it further; our 1992 paper (Abbott *et al.*) was a summary of this second phase of development. We have made further improvements since then, which have been incorporated into the current protocol.

Logic of the Method

The aim is to prepare microvessel fragments from fresh brain, in such a way that contaminating cell types are removed as far as possible, either during the initial preparation, or by selective treatment after the plating of the culture. We wish to isolate microvessels, not larger vessels, firstly because microvessels will have less smooth muscle contamination, and secondly because they are the major site of blood-brain exchange, and hence the most critical location for the blood-brain barrier, the structure of interest. There is also the possibility that large and small vessels will have endothelia of different phenotype; if the aim is to make confluent monolayers in which cells form tight junctions with their neighbours, it may be important to start with a homogeneous microvascular endothelial population.

Vessel fragments are easy to identify, while single cells are not; vessel fragments therefore have an advantage as a starting material for the culture. Furthermore, it turns out that single cells do not grow well in culture, possibly because they lack 'survival factors' from their neighbours. For high yield pure cultures, it is therefore best to aim for good clean vessel fragments. Our method starts with coarse trituration of isolated brain, has two enzymic incubations to free microvessels from contaminating cells, and finishes with a Percoll gradient to isolate the vessel fragments.

The protocol is derived from our working 'recipe' as used in the laboratory. It is deliberately detailed, with everything spelled out, to help new workers learn the method and understand the underlying principles.

MATERIALS AND METHODS

Equipment and Chemicals Required

Laboratory equipment

Denley BR 401 refrigerated centrifuge, max speed 6000 rpm, rotors for Universal tubes, square and round cushions.
Dupont 10 ml ultracentrifuge tubes (03930 + 03279 lids + 00425 converters).
Dissecting instruments set A : (Decapitation, dissection of brain)
Dissecting instruments set B : (Fine dissection, removal of meninges, separation of grey matter from white)
Glassware : beakers, bottles, conical flasks.
Disposable filters, plastics. Sterile Pasteur pipettes, back end stuffed with cotton as filter, some with tips narrowed in flame.

Stock solutions (for composition see Table 1)

Make up in washed and rinsed glassware, all in distilled water (except where marked), then filter through 0.22 μm filter. Dispense into 4–5 ml aliquots except where marked and freeze at $-20\,^{\circ}$C.

TABLE 1 Stock solutions (S = supplement, see section below).

	Label	Use	Final conc. (/ml)	Aliquot size
Penicillin G 300 mg			100 U	
Streptomycin 500 mg	P/S	1:100	100 μg	
both in 50 ml				
TLCK 2.94 mg in 20 ml	T	1:1000	0.147 μg	
Glutamine, 1.46 g in 50 ml	G	1:100	2 mM	
ECGS 15 mg in 2 ml	E	1:100	75 μg	10 × 200 μl
Heparin 400 mg in 50 ml HBSS	H	1:100	80 μg	
Amphotericin B,	F	1:100	2.5 μg	
(Fungizone) 5 mg in 20 ml				
DNAse 1 vial in 1 ml (2000 U)	DN	1:100	20 U	5 × 200 μl
Vitamin C 100 mg in 10 ml	VC	used in S		
Selenium 500 μg/ml	Se	used in S	(**beware** this is highly toxic)	
Basic enzyme mix:	CD	1:1	1 mg	5 × 20 ml
Collagenase/Dispase				
100 mg in 100 ml HBSS (0:Ca;0:Mg)				
+ 0.25 g HEPES (10 mM) + P/S 1 ml				
Collagen *(Sigma) 50 mg in 150 ml		1:1	0.33 mg	
add 2–3 drops glacial acetic acid				
(helps collagen to dissolve)				

*NB: Home-made rat-tail collagen (method of Strom & Michalopoulos, 1982) works better for brain endothelial cells.

Carbodiimide/Collagen coating of glass coverslips (Nobles and Abbott, 1994)

1. (a) H_2SO_4 0.1N 1 hr (500 μl/well)
   (b) Suck off the acid, add 500 μl NaOH 0.1N
   (c) Dry coverslips with gauze
2. (a) Prepare (fresh on the day) carbodiimide solution 130 μl/ml (do it in the hood) 1-cyclohexyl-3-(2-morpholinoethyl) carbodiimide metho-p-toluene sulfonate (Adrich).
   (b) Dilute collagen 1/10 in this carbodiimide solution (carbodiimide/C:10)
   (c) 500 μl carbodiimide/C:10 on each coverslip 3 hr
   (d) Wash 3 × with distilled water
   (e) Sterilize coverslips with ethanol 70% (wash × 2)
3. (a) Coat coverslips with C:10 1 hr
   (b) Cross-link collagen with NH_3 vapour 10 min
   (c) Wash 3 × with HBSS.

Isotonic Percoll gradient medium

50 ml Percoll + 5 ml 10 × strength HBSS (with Ca/Mg) plus 45 ml 1 × strength HBSS (with Ca/Mg). Ca and Mg are required to help capillaries to aggregate/clump, desirable on gradient but not at earlier stages of procedure. Adjust pH by eye, swirling the solution all the time, using 1M HCl. Pure Percoll stock can be autoclaved but not when it is made up as isotonic stock.

Addition of growth factors to medium

Heparin inhibits growth and migration of smooth muscle cells, and has a mildly mitogenic effect on endothelial cells, which is synergistic with ECGS.

Increase in growth:	ECGS	Heparin	
(Hughes, pers. comm.)	−	+	+ 17%
	+	−	+ 63%
	+	+	+200%

Vitamin C is a cofactor in collagen synthesis
Supplement (S), use diluted 1:100. To 10 ml distilled H_2O add:

		Final conc
Vitamin C	500 μl of 10 mg/ml solution (VC)	5 μg/ml
Selenium	10 μl of 500 g/ml (Se)	0.005 μg/ml
Insulin	5 mg	5 μg/ml
Transferrin	5 mg	5 μg/ml
Glutathione	325 mg	325 μg/ml

These are mostly trace factors that help growth. Filter 0.22 μm to sterilise.
Amphotericin B (=Fungizone) kills fungus but also slows down growth. Use only if essential.

Detailed Methodology

Animals and yield

The culture starts with 2–(3) rats, and generates sufficient cells for 60 wells of a 96 well plate, or 2–3 × 35 mm culture dishes at equivalent seeding density. We use Lewis rats (female for immunology, otherwise either sex), usually less than 3 months old. For males, 2 months is optimal as otherwise they grow too big, and meninges become harder to remove cleanly.

Prior preparation (day before)

Sterilise instruments and glassware. Working in the hood:
1. **Prepare Buffer A.** 150 ml HBSS 0:Ca; 0:Mg (Hanks Balanced Salt Solution, Ca and Mg free), up to shoulder in clean (non-sterile) 150 ml bottle.
   0.36 g HEPES (10 mM approx; 10–25 mM is OK)
   1.5 ml stock Antibiotics (P/S, Penicillin/Streptomycin)
   pH to 7.3 (by eye against unopened HBSS) with 1 M NaOH
2. **Prepare BSA (bovine serum albumin) stock.** 20 ml Buffer A in clean (non-sterile) 150 ml conical flask, + 5 g BSA (to give 25% BSA final). Method: put buffer in first, scatter BSA powder on surface, don't swish it around or it will congeal into a poorly-soluble mess. Leave to dissolve, may take > 1.5 h, avoid stirring during this time. Filter through 0.8 μm filter, wetting the filter first with Buffer A.
   Use 0.22 μm filter to sterilise first Buffer A into sterile 150 ml bottle, then BSA stock into sterile Universal tube. Always filter Buffer A first as it must not be contaminated with BSA, otherwise it will go frothy. Keep BSA stock in fridge overnight.
3. **Check stock solution availability.** Collagenase/Dispase solution (Code C/D), 20 ml. Degrades extracellular matrix. DNase (DN) 1 aliquot 200 μl (degrades released DNA which is sticky).
   TLCK (T) 20 μl (inhibits clostripain, a non-specific protease contaminant of collagenase)
   DMEM (Dulbecco's Modified Eagles Medium) 10 ml
   Donor Horse serum 2 ml (alternative: bovine plasma derived serum (lab-made, see Abbott *et al.*, 1992; or Sigma))
   Heparin (H) 120 μl
   ECGS (E, Endothelial cell growth supplement) 1 aliquot
   'Supplement' (S, mixture of vitamins, trace elements etc) 120 μl
   Glutamine (G) 120 μl
   Antibiotics (P/S) 120 μl
4. **Book ultracentrifuge, check disposable supplies, culture lab.**

Procedures on the day: start by 9 am, finish by 5 pm

1. **Preparation of bench and instruments.** Have instruments 'A' ready in the room where you will be decapitating the rats, 'B' in hood. In animal room, prepare: 2ft square Benchcote, small cotton wool swabs, instruments in 70% ethanol, ether bottle or CO_2 for anaesthesia, body bag taped by sink.

2. **Preparation of BSA-Buffer.** Oxygenate Buffer A (95% O_2, 5% CO_2) 10 min, via Pasteur pipette, in hood. Add 3 ml BSA stock solution (1 ml/50 ml, final concentration 0.5% BSA) using blue (1 ml) autopipette tip. Readjust pH by eye against unopened HBSS, with about 10 drops 1M NaOH, using plugged Pasteur pipette. Err on the acid side (orangey) if in doubt. Place 20 ml BSA/buffer (about 1 cm depth) in each of 2 sterile beakers (1 × 75 ml, 1 × 100 ml), cover with sterile foil, onto ice; make sure < 20 ml in larger one (for chopped tissue). Leave c 70 ml BSA/buffer for centrifugation steps. Warm enzyme solution (C/D) in water bath 37 °C; also thaw DNase and TLCK.

3. **Initial dissection of brain.** Total dissection takes about 1 hr for two brains. Start in animal room; wear gloves. Swab down benchcote with 70% ethanol. Soak small scissors, forceps, large scissors, small spatula in ethanol, with some small pledgets of cotton wool. Place 1st rat into fresh ether (several squirts on cotton wool in rat box) until fully anaesthetised. (Alternative, Schedule 1 Procedure: use CO_2). Soak head/neck with ethanol, holding body over bag, against side of bench. Approaching from side, cut behind skull to sever neck, hold by ear and complete cut, allowing body to fall into bag. Clean head with ethanol. Make 2 cuts at sides of cerebellum, and using scissors, bend part of the skull upwards so that you can see the back of the cerebellum. Make a single cut towards the nose along the midline of the skull, avoiding digging scissors into brain. Sliding the flat blade of the scissors under the skull and levering against the sides of the head, fold back one side of the skull, so that it breaks along the skull sutures. Repeat with the other side. Gently ease the brain out with small curved spatula, curving the bent part of the spatula over the nose end, and allow it to fall into the small beaker of BSA/buffer. Cover the beaker and put this back on ice. Repeat process with the second brain. Wash blood off instruments before it sets.

4. **Preparation for fine dissection.** Transfer brains in buffer to sterile room, work in hood with front about 8" open. Place sterile glass Petri dish lid upside down on wet gauze to stop it sliding about. Have fine instruments (set B) sterile, lean each on side of Petri dish base (have ethanol beaker handy and flame instruments that may get contaminated). Scalpel with No 10 blade, large spatula, fine curved forceps, medium (meeting along c 1 cm) and large curved forceps. Spread sterile lint swab in dish bottom with forceps. Add buffer to soak lint fully, surface should look flooded. Spread out dry lint beside dish onfoil with instruments for rolling brains.

5. **Fine dissection.** Remove cerebellum, then slice brain in half longitudinally. Return one half to buffer. From remaining half, pick off meninges and choroid plexus (starting from underside, Circle of Willis) — do this thoroughly with

fine forceps and spatula, pulling it off in large sheets if possible. The surface of the brain then looks almost featureless. Take off hind/midbrain without opening brain (pinch off white mass). Roll brain on dry lint and roll around to remove surface leptomeningeal cells. Turn the lint over and replace brain on the wet lint, then roll to moisten all over. At this stage half-brain looks like a mushroom. With medium blunt but meeting curved forceps, open brain up (peel flap forwards), pinch off white matter and striatum; gives better yield if you remove white matter at this stage. Turn over, take off optic nerve. (Takes about 10 min to this stage). Transfer to other buffer beaker (large), chop with scalpel for <1 min until in uniform 2–3 mm bits. Prepare a fresh lint and dissect second half of brain, transfer to same beaker and chop as before. [Set Denley centrifuge to cool, 4 °C takes about 10 min]. Repeat with other brain, into small beaker to start with. Pool all dissected brains in large beaker and chop as added.

6. **First spin.** Transfer suspension into one sterile Universal. Fill balance tube with water by eye for centrifugation in Denley. Load tubes in outer pockets, using square cushions. Spin cells at 4 °C, 5 min, 1500 rpm = 600 g.

7. **1st enzyme digestion, 1 hr.** Oxygenate 20 ml C/D enzyme mix 2–3 min. Readjust pH to 7.3. Add 1 aliquot DNase I (Pasteur pipette some enzyme mix into bijou tube and suck all back) and 20 μl TLCK. Pour off supernatant from spun tubes. Flick cells off bottom of tube about 10 times; hold Universal tube in left palm, tap side at base smartly with forefinger. Add 15 ml enzyme solution (10 ml pipette). Draw up and down about 3 times to break up large lumps, then place in water bath for 1 hr, shake occasionally (every 10 min), 37 °C (about 10.30–11.30 am). Can prepare Dupont tubes and collagen plates now or during 3 hr digestion: see Points 11, 12 below.

8. **Trituration and second spin.** Triturate suspension up and down with Pasteur pipette (long one stuffed) about 2 min until creamy looking (feel no resistance). Then repeat with narrowed tip to Pasteur (heated in flame to reduce, then sterilised), until all grey matter dispersed (about 2 min). Centrifuge again 1500 rpm, 5 min.

9. **Density-dependent centrifugation in BSA.** There will be a sludge of cells remaining in base of tube. Pasteur pipette off supernatant (don't pour), to leave sludge at bottom. Resuspend by shaking tube. Add neat 25% BSA (17 ml = rest of solution prepared earlier). Invert tube to mix, then break up with (rounded) Pasteur tip. Balance tubes very carefully for faster spin, spin 15 min 2900 rpm (1000 g). The theory of this centrifugation is that neurones, astrocytes and myelin will float, capillary fragments will sink through BSA. Single cell suspensions of endothelial cells are not viable and will also float. BSA is preferred to Ficoll, which wrecks cells.

10. **Spin and 2nd enzyme digestion.** Roll tube gently to free myelin from sides of tube. Pour off top layers including myelin to leave capillary pellet, using fairly fast initial swish. Replace cap, leave tube upside down to avoid any myelin sliding back down to contaminate. Resuspend pellet in Buffer A by pipetting small volume into the bottom (swish up and down fast) then transferring to new Universal. Repeat to wash out residue. Some fibres will be present, left

over from lint, no problem. Fill to about half full with Buffer A, spin down again, 1500 rpm, 5 min. Pour off supernatant. Flick to resuspend in remaining 5 ml enzyme digest solution, incubate 3 hr 37 °C, with occasional shaking.

During 3 hr enzymic incubation (approx 12.30–3.30 pm):

11. **Prepare collagen-coated plates.** Use 6 ml collagen solution for ∼100 μl in each of 60 wells of 96 well Nunclon plate, or 2–3 × 35 mm dishes. Leave 1–2 hrs with lid on in hood. Use ammonia solution to cross-link collagen in NH_3 vapour: (a) Add about 3 Pasteur-fulls 35% NH_3 (BDH 19007) onto tissue in plastic box. (b) Using fire-polished Pasteur to avoid scratching surface, remove collagen solution from wells. (c) Place plate in NH_3 box for 10 min, lid still on plate. (d) Allow ammonia vapour to evaporate by removing lid from plate. Place wash medium (HBSS not Buffer A) into wells (will have total 3 washes). After 2nd wash, place in incubator. [The plate can be prepared at any stage; having the plate sitting in the incubator is not a critical step, but don't let collagen dry out as this denatures it, and our endothelial cells don't like denatured collagen].

** *To get cells to stick to glass, need protocol to cross-link collagen to glass before re-coating with fresh collagen. See instructions above (Nobles & Abbott, 1994).*

12. **Prepare culture medium** (fresh each time). Thaw stocks in water bath, 37 °C.
    (a) 10 ml DMEM (Flow)
    (b) 2 ml donor horse serum DHS (Flow) or plasma-derived bovine serum (Sigma). Use same 10 ml pipette for (a) then (b).
    Use same yellow pipette tip for the following, in order: 120 μl glutamine, G
    120 μl heparin, H
    Antibiotics: 120 μl P/S
    120 μl 'Supplement', S
    1 100 μl aliquot ECGS, E
    120 μl amphotericin B (if needed, fungicide; avoid if possible, inhibits endothelial growth).
    Filter 0.22 μm before use. This is enough for 60 wells of 96 well plate (× 200 μl per well, 12 mls total) or 2–3 × 35 mm dishes. For 24 well plates, each well takes 500 μl, filter insert takes 250 μl, about 350 μl outside, so 12 ml is enough for 24 wells without inserts, 20 wells with inserts.

13. **Prepare Dupont ultracentrifuge tubes**, previously cleaned by sonication in Milton. To improve the separation on the gradient, you should fill the tubes with distilled H_2O and resonicate before use. Sterilise now by half-filling with ethanol and shaking to coat fully (about 2 min+, make sure caps/inserts are fully wetted). Tip out ethanol, wash with about 2 ml buffer, dump this. Add 8 ml new buffer, stir on shaker or rotator 20 min + to coat walls with protein to prevent cell adhesion to plastic.

14. **Make Percoll gradients.** Use prepared Percoll gradient medium (14 ml isotonic stock). Swirl to mix before pipetting. Using prepared 10 ml Dupont ultracentrifuge tubes, carefully take off caps, put them down so they stay sterile

and spacers don't fall out. Pipette 7 ml of isotonic Percoll stock into each tube. Tighten caps over spacers/gaskets. Place inside grey adaptors in Sorvall centrifuge, tighten head lid lower nut then upper nut, anticlockwise. Form gradients by centrifugation for 1 hr 25,000 g. (e.g. Sorvall RC-2B, floating head, 16,500 rpm 4 °C; set to TIMER, press start, set time, speed up. Alternative: Sorvall RC5B or MSE Europa fixed head 34 ° angle). Centrifugation may produce a band of aggregated Percoll, this is not a serious problem.

After 3 hr digestion:

15. **Percoll gradient separation of capillary fragments.** Before spinning cells, get rid of unwanted debris in digest: draw digest up into Pasteur pipette held almost horizontal, allow fibres and debris to settle, squeeze bulb gently to expel capillaries not fibres (fibres stick to glass, capillaries don't). Spin enzyme digest 1,500 rpm, 5 min. Pour off supernatant, leaving pellet. Resuspend pellet by tapping tube. Take roughly 1 ml Buffer A (1 Pasteur-pipette-full), resuspend pellet. Stand Pasteur pipette upright in suspension. Holding at an angle, withdraw Dupont tube slightly from adaptor, take off cap and prop up so stays sterile and doesn't lose spacer. Suck up half suspension, gently dribble onto top of one Percoll gradient, holding tube at an angle. Repeat for second tube. Using round bottom rubber cushions in the Denley centrifuge, spin Dupont tubes 10 min 2900 rpm (1000 g) to separate capillary fragments from contaminating cells.

16. **Preparation of culture plates and medium.** Wash plates finally in HBSS; using fire-polished pipette so as not to scratch collagen surface, suck off previous wash from plate wells and add new HBSS. Prepare 6–7 ml (bijou tube), half Buffer A/half DMEM (not full culture medium yet, reduces osmotic shock caused by solution change). This wash is used to wash Percoll off cells.

17. **Collection of capillary fragments.** Prepare Universal tube with 10 ml Buffer A. Stand Pasteur pipette in it. Kneel down to see properly, holding first Dupont tube at eye level in hood. Take pipette rinsed in Buffer A and squeeze out. Start with bubble in Pasteur pipette, and applying positive pressure to avoid sucking in upper layers, lower pipette tip to capillary layer (frothy/particulate, about 4/5ths way down, just above layer of any Percoll aggregates). Holding fingers behind tube to see layer clearly, suck up about 0.5–1 ml from capillary region of gradient. Add to 10 ml prepared Buffer A. Repeat with second Dupont tube. Balance Universal tubes, spin 1800–1900 rpm, 5 min. Pour off supernatant. Tap tube to resuspend cells. Add half and half Buffer A/DMEM. Respin 5 min, 1500 rpm. Suspend finally in 12 ml prepared culture medium, invert tubes a few times to mix well.

18. **Plating cells.** Remove final wash from plate 1 row at a time, using fire-polished long Pasteur. Add 200 μl suspended cells in medium to each well (96 well plate; adjust for other sizes), using blue pipette tip. Change blue tip between each row to avoid contamination. Agitate the suspension to ensure even distribution of capillaries between wells. Use central wells of plate not edge wells (more variable conditions in incubator).

19. **Feeding, contaminants.** Cultures are fed Mon, Wed, Fri: suck off medium a row at a time, and add fresh medium. Cells can be washed after 1–2 day with Ca- and Mg-free HBSS to remove contaminants. Cells become confluent after 7–10 days.

20. **Complement killing of contaminating pericytes.** At 3–4 days after isolation, or 1–2 days after subculturing if cells are passaged.
    (a) Wash dishes twice with medium without serum.
    (b) Incubate in polyclonal anti-Thy 1.1. antiserum (1:500 dilution in medium without serum) for 40 min at 37 °C in the incubator (5% CO_2 atmosphere).
    (c) Wash dishes twice with medium without serum.
    (d) Incubate with rabbit complement for 120 min at 37 °C in incubator.

The complement (Sigma) is reconstituted in 5 ml water 30 min then 2 ml medium without serum added. 100–200 μl of this stock is added to each well. Complement can be reconstituted (water/medium) and kept at –70 °C for 1–2 months.

(e) Stop the killing process when pericytes and astrocytes are lysed; wash dishes twice with medium without serum and feed with normal medium with supplements.

SPECIFIC APPLICATIONS

The primary rat culture has been used in our lab for a number of studies, on transport (cyclosporin, Begley *et al.*, 1990; albumin, Ramlakhan, 1990), receptor-mediated changes in intracellular calcium (Revest *et al.*, 1991; Nobles *et al.*, 1995), toxicology (Romero *et al.*, 1992), actin cytoskeleton (Rist *et al.*, 1994) and fine structure (Lane *et al.*, 1995).

PROBLEM SOURCES AND QUALITY CONTROL

Using this procedure, cultures that are >95% pure endothelial can be produced, which is good enough for most studies of receptors, transporters and biochemistry. Pericyte contamination proves to be a greater problem for rat cultures than for bovine; in the former, each pericyte causes a small flaw in the monolayer, which will act as a leak however tight the junctions between neighbouring endothelial cells. In bovine cultures, the endothelium appears better able to seal over the places occupied by pericytes, so tighter layers can be achieved; moreover, it appears to be easier to passage pure clones from bovine endothelial plaques without pericytes and to subculture these. Nevertheless, rat cultures have been made with up to ~150–200 Ω.cm^2 transendothelial resistance, which is around $100 \times$ tighter than mesenteric capillaries *in vivo*, and tight enough to address many kinds of experimental question.
    The cells preserve markers for endothelia (FVIII-related antigen, angiotensin-converting enzyme, binding of lectin from *Ulex europaeus*, tight junctions), and some 'barrier phenotype' markers (transferrin receptor, p-glycoprotein). Further

barrier features lost in culturing can be re-induced by co-culture with astrocytes or by addition of astrocyte-conditioned medium and elevation of intracellular cAMP (marginal actin, γ-glutamyl transpeptidase). In these respects the rat model resembles the bovine preparation, and is a basis for developing an *in vitro* BBB.

The advantages of the rat culture are its suitability for comparisons with rat studies *in vivo*, the availability of Thy 1.1 antibody for contaminant killing, and the availability of other antibodies for identifying blood-brain barrier 'marker' characteristics. The disadvantages are the low yield of capillary fragments from such small brains, and the effects of pericytes and other contaminants on the permeability of the monolayer.

CONCLUDING REMARKS

This culture method has now been in use for nearly 10 years, and cells produced by the method have been used for a variety of studies on the physiology, pharmacology, and immunology of the brain endothelium. This literature is itself a valuable reference resource, so that further studies on brain endothelial cells produced by the method can add to the accumulating information and build on it, reducing the effort required in developing a method from scratch and characterising the cells that result. The method has been successfully learnt by ~ 10 research staff in our group, and transferred to other laboratories, where the results are consistent and reproducible. This shows that the method is robust, practical and reliable.

ACKNOWLEDGEMENTS

Establishment of this method at King's College was supported by Cambridge NeuroScience Inc (USA), and subsequent work characterising the cells and using them for research projects was funded by the MRC, Wellcome Trust and Royal Society. We have benefited from technical discussions at Brain Endothelial Workshops organised under the Biomed-1 Concerted Action programme, and are grateful to Dr Bert de Boer and Dr Win Sutanto for organising these meetings.

REFERENCES

Abbott, N.J., Hughes, C.C.W., Revest, P.A. and Greenwood, J. (1992) Development and characterisation of a rat brain capillary endothelial culture: towards an *in vitro* blood-brain barrier. *J. Cell Sci.*, **103**, 23–37.

Begley, D.J., Squires, L.K., Zlokovic, B.V., Mitrovic, D.M., Hughes, C.C.W., Revest, P.A. and Greenwood, J. (1990) Permeability of the blood-brain barrier to the immunosuppressive cyclic peptide cyclosporin A. *J. Neurochem.* **55**, 1222–1230.

Bowman, P.D., Betz, A.L., Wolinsky, J.S., Penny, J.B., Shivers, R.R. and Goldstein, G.W. (1982) Primary culture of capillary endothelium from rat brain. *In Vitro* **17**, 353–362.

Bowman, P.D., Ennis, S.R., Rarey, K.E., Betz, A.L. and Goldstein, G.W. (1983) Brain microvessel endothelial cells in tissue culture : a model for study of blood-brain barrier permeability. *Ann. Neur.*, **14**, 396–402.

Greenwood, J., Adu, J., Davey, A.L., Abbott, N.J. and Bradbury, M.W.B. (1991) The effect of bile salts upon the permeability and ultrastructure of the perfused, energy-depleted, rat blood-brain barrier. *J. Cerebr. Blood Flow Metab.*, **11**, 644–654.

Hughes, C.C.W. and Lantos, P.L. (1986) Brain capillary endothelial cells lack surface IgG Fc receptors. *Neurosci. Lett.*, **68**, 100–106.

Hughes, C.C.W. and Lantos, P.L. (1989) Uptake of leucine and alanine by cultured cerebral capillary endothelial cells. *Brain Res.*, **480**, 126–132.

Lane, N.J., Revest, P.A., Whytock, S. and Abbott, N.J. (1995) Fine-structural investigation of rat brain microvascular endothelial cells: tight junctions and vesicular structures in freshly isolated and cultured preparations. *J. Neurocytol.*, **24**, 347–360.

Nobles, M. and Abbott, N.J. (1994) A method for growing brain endothelial cells on glass. *J. Physiol.*, **480**, 4–5P.

Nobles, M., Revest, P.A., Couraud, P.-O. and Abbott, N.J (1995) Charcteristics of nucleotide receptors that cause elevation of cytoplasmic calcium in immortalized rat brain endothelial cells (RBE4) and in primary cultures. *Br. J. Pharmacol.*, **15**, 1245–1252.

Ramlakhan, N. (1990) Albumin binding and endocytosis by cultured rat brain endothelium. *J. Physiol.*, **423**, 34P.

Revest, P.A., Abbott, N.J. and Gillespie, J.I .(1991) Receptor-mediated changes in intracellular [Ca^{2+}] in cultured rat brain capillary endothelial cells. *Brain Res.*, **549**, 159–161.

Rist, R.J., Romero, I.A. and Abbott, N.J. (1994) The effects of a cAMP analogue and astrocyte-conditioned medium on the F-actin cytoskeleton in cultured primary and immortalized rat brain capillary endothelial cells. *J. Physiol.*, **480**, 8–9P.

Romero, I.A., Cavanagh, J.B., Nolan, C.C., Ray, D.E. and Seville, M.P. (1992) 1,3-dinitrobenzene, a neurotoxin acting at the rat blood-brain barrier. *J. Physiol.*, **446**, 498P.

Rubin, L.L., Hall, D.E., Porter, S. *et al..*, (1991) A cell culture model of the blood-brain barrier. *J. Cell Biol.*, **115**, 1725–1735.

Strom, S.C. & Michalopoulos, G. (1982) Collagen as a substrate for cell growth and differentiation. *Methods in Enzymology*, **82**, 544–548.

II.2. POSSIBILITIES, LIMITATIONS AND ISOLATION PROCEDURES OF RAT BRAIN MICROVESSEL ENDOTHELIAL CELL (RMEC) CULTURE TECHNIQUES

A.G. de BOER, H.E. de VRIES, P.J. GAILLARD AND D.D. BREIMER

Division of Pharmacology, Leiden/Amsterdam Center for Drug Research (LACDR), University of Leiden, P.O. Box 9503, 2300 RA Leiden, The Netherlands

In this chapter, the use of brain microvessels, and the isolation and culture of rat brain microvessel endothelial cells are described. Pitfalls in the application and culture of these cells are also explained. In addition, the isolation and culture of astrocytes and the collection of astrocyte-conditioned medium is described.

INTRODUCTION

Research on blood-brain barrier (BBB) transport of drugs has been considerably facilitated and enhanced by the introduction of *in vivo* and *in vitro* techniques. The *in vivo* techniques (van Bree *et al.*, 1992; de Lange *et al.*, 1995) allow the estimation of the rates and quantification of drugs transported across the BBB, while the *in vitro* techniques allow a more detailed study, qualitatively as well as quantitatively, of BBB transcytosis processes. However, the results obtained by *in vitro* and *in vivo* techniques can differ considerably, which hinder the extrapolation of *in vivo* results from the *in vitro* situation. The major causes of these discrepancies are due to the properties of the various *in vitro* and *in vivo* models used. In this chapter the possibilities and limitations of some of the *in vitro* BBB models will be discussed. The materials and procedures for the isolation of rat brain microvessel endothelial cells (RMEC) are presented. The procedures for bovine brain microvessel endothelial cells (BMEC) are described in section III.1.

POSSIBILITIES AND LIMITATIONS

The *in vitro* methods comprise the application of isolated brain microvessels (Joó, 1992) and the isolation and culture of brain microvessel endothelial cells (BMEC) (van Bree *et al.*, 1992; Joó, 1992; Balconi and Dejana, 1986; Pardridge, 1991).

Isolated Brain Microvessels

Brain microvessels have been used from the very beginning of BBB research (Joó, 1992; Pardridge, 1991; Joó, 1985). They are well suited for studying the uptake of nutrients and localizing or identifying specific enzymes and substrate binding sites. In addition, isolated brain microvessels have been used in the field of drug disposition (uptake into endothelial cells) and metabolism.

The major advantage of the microvessel isolation technique is that it provides a system that is very similar to the *in vivo* situation with respect to the expression of surface molecules and might therefore be a good screening model to identify selective or specific interactions with the microvessel endothelium. However, this technique suffers from a number of drawbacks:

- it is a heterogenous system (White *et al.*, 1992);
- it is not suitable for studying transcytosis processes (Joó, 1985);
- in uptake studies the substrate approaches the endothelial cells mainly from the abluminal side (Lidinsky and Drewes, 1983);
- the energy supply of the endothelial cells is limited which is a serious drawback in energy dependent uptake studies (Joó, 1985).

Isolation and Culture of Brain Microvessel Endothelial Cells (MEC)

The isolation and culture of MEC has tremendously enhanced research in the area of drug transport across the BBB, in particular at the (sub)cellular level. The availability of cultures on porous membranes or filters has contributed much to the applicability of this system in providing the growth of confluent monolayers of MECs. In addition, co-culture of MEC with astrocytes has enhanced the functional properties of the *in vitro* BBB and increased expression of various endothelial markers such as gamma-glutamyl transpeptidase (γ-GTP), monoamine oxidase (MAO), angiotensin converting enzyme (ACE). These systems are useful for studying a variety of phenomena, e.g. drug transport across the BBB; drug metabolism; the effect of agents on the functionality of the BBB; etc.

Various techniques have been applied for the isolation and culture of MECs (Joó, 1992; Balconi and Dejana, 1986; Pardridge, 1991; Joó, 1985; van Bree *et al.*, 1988). These comprise, following dispersion of the collected grey matter from the brain, the application of enzymatic (Audus and Borchardt, 1986) or combined mechanical and enzymatic (Rubin *et al.*, 1991) procedures. It results in the isolation and culture of MEC that may differ in various properties, such as the expression of Von Willebrand Factor (VWF), MAO, γ-GTP, uptake of acetylated low density lipid particles (DiI-AcLDL), differences in trans-endothelial electrical resistance (TEER), and differences in (para- and transcellular) permeability. The main causes responsible for this variability are listed below:

- Species differences for the source for brain microvessel endothelial cells (MEC): rat (Rubin *et al.*, 1991), bovine (van Bree *et al.*, 1988), calf (van Bree *et al.*, 1988), pig (Mischeck *et al.*, 1989), in particular with respect to the number of viable cells isolated (Mosquera and Goldman, 1991). In addition, it is our experience that MEC isolated from calf brain grow better than those from cow brain.

- The isolation technique to obtain brain microvessel endothelial cells: mechanical dispersion of the grey matter of the brain followed by enzymatic (Rubin *et al.*, 1991; van Bree *et al.*, 1988) or combined mechanical/enzymatic procedures (Rubin *et al.*, 1991; Pardridge *et al.*, 1986).

- The selectivity of the isolation techniques used with respect to contaminating cell types (Kumar *et al.*, 1989) e.g. pericytes, fibroblasts, and the presence of venous endothelial cells: contaminating cells may influence the TEER, transport through the monolayer, influence the quantitative aspects of drug transport and may lead to erroneous conclusions with respect to the expression of transcellular transport systems and/or surface molecules.

- The use of primary cell cultures (Rubin *et al.*, 1991) vs. passaged cells (Méresse *et al.*, 1989): primary cells mostly have preserved their properties to a certain extent, while passaged cells do, or might have, considerably increased or decreased expression of surface molecules (Albelda, 1991).

- The filter material and the applied extracellular matrix (ECM) where the cells are grown on: it has been shown that the expression of surface molecules is changed depending on the type of extracelluar matrix applied (Dejana *et al.*, 1988). In addition the ECM, hyaluronic acid and also some growth factors seem to be important with respect to the occurrence of angiogenesis (Townsend *et al.*, 1991; Kumar *et al.*, 1989). Components of the ECM, for example, fibronectin and laminin have shown to convert quiescent astrocytes into proliferating ones (Nagano *et al.*, 1993).

- The culture medium applied (serum vs. serum-free medium): serum free medium is well defined although it may lack certain (essential?) and unknown compounds which are present in serum containing medium.

- The application of astrocyte-conditioned medium: the functionality of MEC is, in terms of the expression of surface molecules (VWF, MAO, γ-GTP), *in vitro*, very much dependent on the application of astrocyte-conditioned medium (Rubin *et al.*, 1991; Meyer *et al.*, 1991; Pardridge *et al.*, 1990).

- The co-culture of brain-microvessel endothelial cells with astrocytes: simultaneous growth of MEC and astrocytes provide also good conditions to obtain MEC with good functionality (Albelda, 1991; Janzer and Raff, 1987). In addition, human foetal astrocytes have shown to induce the expression of factor VIII, the GLUT-1 glucose transporter and γ-GTP on autologous endothelial cells (Hurwitz *et al.*, 1993).

- The application of stimulants to induce certain specific features in confluent monolayers of brain microvessel endothelial cells: application of phosphodiesterase inhibitors and compounds that increase intracellular cAMP and as a result the tightness of tight-junctions (Rubin *et al.*, 1991).

- The culture of MEC under static vs. flow conditions.

- Cloning techniques that allow the culture of MEC up to passage 5–10 (Méresse *et al.*, 1989).

- The application of cell lines e.g. the immortalized rat brain endothelial cell line RBE4 (Roux *et al.*, 1994).

Conclusions

All these aspects may influence, to a greater or lesser extent, the experimental results obtained. This may also lead one to conclude that a complete artificial BBB system should be developed, comprising the presence of endothelial cells, astrocytes, pericytes, a basement membrane and neurons. However, this would be a very complicated system and therefore impractical. The most pragmatic solution to this problem might be to look for that particular BBB model that permits one to study a particular aspect of drug transport across the BBB. Systems such as those reported by Méresse *et al.* (1989), Abbott *et al.* (see Part A, II.1), Risau *et al.* (1990) and Rubin *et al.* (1991) may be considered as examples of this pragmatic approach. In addition, before using such systems, they have to be characterized very well with respect to the (transport)-system to be studied. Once these *in vitro* studies have been performed, *in vivo* studies should immediately be carried out.

ISOLATION AND PRIMARY CULTURES OF RAT BRAIN MICROVESSEL ENDOTHELIAL CELLS (RMEC) (Modified from Risau *et al.*, 1990, and Rubin *et al.*, 1991)

Reagents and Materials

- 10 rat cerebral cortices
- Tissue culture petri dishes (100 nm, 35 mm Falcon)
- Surgical instruments including blades
- Centrifuge
- Ultracentrifuge
- Waterbath
- Buffer A: 153 mM NaCl, 5.6 mM KCl, 2.3 mM CaCl$_2$, 15 mM HEPES, 10 mg/ml BSA $\Longrightarrow$ pH 7.4
- Collagenase "Worthington", (CLS II, Nr. C11-22, Biochrom KG) $\Longrightarrow$ 0.75% (w/v) in PBS
- BSA (Albumin fraction V 735108, Boehringer, Mannheim): $\Longrightarrow$ 25% (w/v) BSA in buffer A
- Dulbecco's modified Eagles's medium (DMEM, high glucose formula, Gibco 430–1600)
- DMEM/10% FBS
- DMEM 10-fold (Gibco, 042–2501)
- Percoll (density 1.132 g/ml, code 17–0891–01, Pharmacia) $\Longrightarrow$ 50% (w/v) Percoll: 22.5 ml Percoll + 2.5 ml DMEM 10-fold + 25 ml DMEM/10% FBS
- Collagen (type I and III, Biochem KG) $\Longrightarrow$ dilute 1:1 with PBS for coating of 35 mm petri dishes (1 h, 37°C)

- Trypsin/EDTA: 0.03% trypsin/0.02% EDTA (Gibco)
- RMEC-Medium: DMEM supplemented with 10% calf serum, 1% bovine retinal extract as growth factor (D'Amore and Klagsbrun, 1984), 100 U/ml penicillin, 100 μg/ml streptomycin, 1% non-essential amino acids, sodium pyruvate (1 mM), L-glutamine (2 mM), 0.5 μM 2-mercaptoethanol
- Dimethylsulphoxide (DMSO)

Isolation Procedure

1. Take 10 rat cerebral cortices.
2. Wash cortices in buffer A.
3. Dissect free of meninges and white matter in buffer A.
4. Rinse tissue with surgical blades until resuspendable with 10 ml pipettes.
5. Transfer in 50 ml tube and centrifuge (5 min, 250 g, 20 °C).
6. Resuspend pellet in 0.37% (w/v) collagenase (= 5 ml buffer A + 5 ml 0.75% (w/v) collagenase).
7. Digest for 1.5 h at 37 °C in a water bath, shake gently every 15 min.
8. Stop digestion by adding 40 ml buffer A.
9. Pellet the tissue (250 g, 5 min, 4 °C).
10. Resuspend the pellet in 40 ml 25% (w/v) BSA.
11. BSA gradient: 1000 g, 20 min, 4 °C); banding of the gradient.

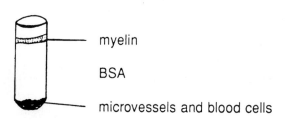

12. Transfer myelin and BSA into a new 50 ml tube, resuspend pellet and run a second BSA gradient.
13. Resuspend the remaining pellet containing the microvessels in 5 ml of 0.75% (w/v) collagenase and transfer into a new tube.
14. Discard myelin and BSA from the second BSA gradient and resuspend the pellet in collagenase as above; keep the pellets separated, because usually the second is contaminated to a varying degree with non-endothelial cells.
15. Digest both pellets for 1.5 h in the water bath at 37 °C, resuspend for several times with a pasteur pipette.
16 Establish Percoll gradient by centrifugation of 50% (v/v) Percoll in 15–20 ml tubes at 25000 g and 20 °C for 1 h in a swing-out rotor.
17. Stop digestion by adding 10 ml DMEM/10% FBS.
18. Centrifuge for 5 min, 250 g, 20 °C.
19. Resuspend each suspension in 1 ml DMEM/10% FBS.

20. Layer each pellet onto the Percoll gradient.
21. Band the microvessels by centrifugation (10 min, 1000 g, 20 °C); banding of the Percoll gradient.

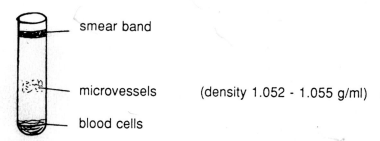

microvessels   (density 1.052 - 1.055 g/ml)

22. Remove each microvessel band and wash twice with DMEM/10% FBS (5 min, 250 g, 20 °C).
23. Resuspend in 6 ml RMEC medium.
24. Plate each suspension onto 3 collagen-coated 35 mm petri dishes.
25. Allow microvessel fragments to attach for 2 h in an incubator (37 °C, 10% CO_2).
26. Remove supernatant and plate onto fresh collagen-coated 35 mm Petri dishes.
27. Add new RMEC-medium to the first RMEC cultures.
28. After an overnight incubation wash with DMEM/10% FBS and add new medium.
29. Fed cells every 3–4 days.
30. An optimal purity of RMEC cell cultures is obtained by selectively lysing contaminating cells such as pericytes and astrocytes on day 5 of the culture (see **Pericyte killing**).
31. Check purity of cultures by immunofluorescence staining for factor VIII-related antigen.
32. RMECs reach confluency after about 2 weeks.
33. To passage them wash RMEC monolayers with PBS; add 0.5 ml trypsin/EDTA and shake gently.
34. Stop trypsinization with DMEM/10% FBS and pellet the cells (5 min, 250 g, 20 °C).
35. For further culturing resuspend in RMEC medium and split 1:3.
36. For freezing, resuspend RMECs of two 35 mm petri dishes in 0.5 ml DMEM/20% FBS + 10% DMSO.

Pericyte Killing Procedure (Risau _et al._, 1990)

Antiserum to Thy 1.1

Materials

- surgical instruments
- homogenizer or stainless steel mesh

- AKR/J mice (Thy 1.1)
- AKR/Cum or C3H mice (Thy 1.2)
- Eagles HEPES (EH)
- 1 ml syringes

Method

- prepare thymocyte suspension from AKR/J donors by homogenizing the thymi or by passing them through a stainless steel mesh
- wash twice with EH
- inject $3 \cdot 10^7$ thymocytes intraperitoneally into each C3H or AKR/Cum recipient weekly for 10 weeks
- bleed immunized mice in week 11 and pool serum
- inactivate serum (56 °C, 30 min)
- filter sterile and freeze (-20 °C)

Pericyte killing

Materials

- DMEM (high glucose formula, Gibco)
- anti-Thy 1.1 antiserum
- rabbit complement (ORAX 06/07, Behringwerke AG, FRG)

Method

- wash cultures twice with DMEM
- add Thy 1.1 antiserum (dilution 1:50 to 1:100, depending on the cytotoxic capacity of each antiserum), final volume 500 μl/35 mm Petri dish
- incubate for 40 min (37 °C, 10% CO_2)
- wash twice with DMEM
- add rabbit complement (dilution 1:10); final volume, 500 μl/35 mm Petri dish
- incubate at 37 °C, 10% CO_2 atmosphere
- check cultures every 10 min and stop the procedure immediately when all contaminating cells are lysed
- wash twice with DMEM
- add fresh RMEC-medium

Isolation and Culture of Astrocytes and the Collection of Astrocyte Conditioned Medium (ACM)

Astrocytes can be isolated from rat pups according to the procedure of Tio *et al.* (1990). Oligodendrocytes are removed after 8 days of culture (McCarty and De Vellis, 1980), and subsequently the astrocyte conditioned medium (ACM) is collected during further culture. The astrocytes are characterized by their glial fibrillary acidic protein (GFAP) expression and are in our hands pure to at least 95%.

(All instruments used must be autoclaved)

1. Use 6 rat pups (2 pups/flask), of one/two days old.
2. Transport them in a clean petri dish instead of in their cage (infections!).
3. Dip the rats in 70% alcohol and lay them in a petridish.
4. Decapitate the rats.
5. Remove the brains and transport them to a sterile petri dish, containing 10 ml of cold DMEM buffer.
6. Isolate the cortex with sterile forceps and remove the meninges. Cut the cortex in pieces and put them in a 50 ml Falcon tube with 10 ml DMEM buffer, 37 °C.
7. Add 5 ml trypsin-EDTA, 37 °C (Gibco) to the 10 ml suspension resulting in a final concentration of 0.1% trypsin.
8. Incubate in a shaking waterbath (80 rpm, 37 °C for 25 min).
9. Add DMEM + 10% FCS to stop trypsinization to a total volume of 50 ml.
10. Spin the cell suspension for 5–8 min at 250 g.
11. Resuspend the pellet in 10 ml DMEM + 10% FCS, then filter the cell suspension through a 120 μm mesh and rinse with 10 ml DMEM + 10% FCS.
12. Filter the cell suspension through a 45 μm mesh and rinse with 10 ml DMEM + 10% FCS.
13. Seed the cells in flasks (T75 Falcon flask), 10 ml DMEM + 10% FCS, 37 °C each.
14. Culture at 37 °C, 5% CO_2 for approx. 8 days, but refresh the DMEM + 10% FCS after three days (not before!).
15. When confluent (after 8 days) shake the culture flasks in a dry and clean (0.1% Triton; 70% ethanol) shaking waterbath at the highest speed (250 rpm) overnight, then refresh the medium (10 ml DMEM + 10% FCS). In this way the type 1 astrocytes are separated from type 2, oligos and microglia, because astrocytes are firmly adhered to the culture flask while the oligos and microglia are not.
16. Culture at 37 °C, 5% CO_2 until confluence, then wash the cells with 10 ml PBS (37 °C).
17. Passage with 5 ml trypsin-EDTA (Gibco), 37 °C for approx. 10 min (shake and tap until most of the cells detach).
18. Pool and add 35 ml DMEM + 10% FCS to stop trypsinization.
19. Spin the cell suspension for 5 min at 125 g.
20. Resuspend the pellet in 90 ml DMEM + 10% FCS and plate the cells in nine poly-D-lysine-coated flasks, 10 ml each.
21. Change the medium every other day until confluence and then collect the astrocyte conditioned medium every 48 h for maximal 6 weeks and store in 50 ml tubes at −20 °C.

Coating of the culture flasks for the astrocytes

1. Dilute the poly-d-lysine stock solution 100 times (from 1 mg/ml to 10 μg/ml) with Milli-Q water (sterile filtered, 0.2 μm) or use the already diluted solution.

2. Add 6 ml per culture flask (T75) and leave overnight on a shaker at room temperature.
3. Aspirate and allow to dry.
4. Wash with PBS buffer (3×) before plating the cells.

REFERENCES

Albelda, S.M. (1991) Endothelial and epithelial cell adhesion molecules, *Am. J. Respir. Cell. Mol. Biol.*, **4**, 195–203.
Audus, K.L., Borchardt, R.T. (1986) Characterization of an *in vitro* blood-brain barrier model system for studying drug transport and metabolism, *Pharm. Res.*, **3**, 81–87.
Balconi, G. and Dejana, E. (1986) Cultivation of endothelial cells: limitations and perspectives, *Medical Biol.*, **64**, 231–245.
D'Amore, P.A. and Klagsburm, M. (1984) Endothelial cells mitogens derived from retina and hypothalamus: biochemical and biological similarities, *J. Cell Biol.*, **99**, 1545–1549.
De Lange, E.C.M., Hesselink, M.B., Danhof, M., De Boer, A.G., and Breimer, D.D. (1995) The use of intracerebral microdialysis to determine changes in blood-brain barrier transport characteristics, *Pharm. Res.*, **12**, 129–133.
Dejana, E., Colella, S., Conforti, G., Abbadini, M., Gaboli, M., and Marchisio, P.C. (1988) Fibronectin and vitronectin regulate the organization of their respective Arg-Gly-Asp adhesion receptors in cultured human endothelial cells, *The Journal of Cell. Biol.*, **107**, 1215–1223.
Hurwitz, A.A., Berman, J.W., Rashbaum, W.K. and Lyman, W.D. (1993) Human fetal astrocytes induce the expression of blood-brain barrier specific proteins by autologous endothelial cells, *Brain Res.*, **625**, 238–243.
Janzer, R.C. and Raff, M.C. (1987) Astrocytes induce blood-brain barrier properties in endothelial cells, *Nature*, **325**, 253–257.
Joó, F. (1992) The Cerebral Microvessels in Culture, an Update, *J. of Neurochemistry*, **58**, 1–17.
Joó, F. (1985) The blood-brain barrier *in vitro*: ten years of research on microvessels isolated from the brain, *Neurochem. Int.*, **7**, 1–25.
Kumar, S., Kumar, P., Pey, D., Sattar, A., Wang, M., and Ponting, J. (1989a) Heterogeneity in endothelial cells with special reference to their growth related proteins, In: *Angiogenesis in Health and Diseases*, Plenum Press, New York, pp. 63-78.
Kumar, S., Kumar, P., Ponting, J.M., Sattar, A., Rooney, P., Pey, D. and Hunter, R.D. (1989b) Hyaluronic acid promotes and inhibits angiogenesis, In: *Angiogenesis in Health and Diseases*, Plenum Press, New York, pp 253–263.
Lidinsky, W.A. and Drewes, L.R. (1983) Characterization of the blood-brain barrier: protein composition of the capillary endothelial cell membrane, *J. Neurochem.*, **41**, 1341–1348.
McCarthy, K.D. and De Vellis, J. (1980) Preparation of separate astroglial and oligodendroglial cultures from rat cerebral tissue. *J. Cell Biol.*, **85**, 890–902.
Méresse, S., Dehouck, M-P., Delorme, P., Bensaïd, M., Tauber, J-P., Delbart, C., Fruchart, J-C., and Cecchelli, R. (1989) Bovine brain endothelial cells express tight junctions and monoamine oxidase activity in long-term culture, *J. of Neurochem.*, **53**, 1363–1371.
Meyer, J., Rauh, J. and Galla, H-J. (1991) The susceptibility of cerebral endothelial cells to astroglial induction of blood-brain barrier enzymes depends on their proliferative state, *J. of Neurochem.*, **57**, 1971–1977.
Mischeck, U., Meyer, J., and Galla, H.J. (1989) Characterization of gamma-glutamyl transpeptidase activity of cultured endothelial cells from porcine brain capillaries, *Cell. Tiss. Res.*, **256**, 221–226.
Mosquera, D.A. and Goldman, M. (1991) Endothelial cell seeding, *Br. J. Surg.*, **87**, 655–660.
Nagano, N., Aoyagi. M. and Hirakawa, K. (1993) Extracellular matrix modulates the proliferation of rat astrocytes in serum-free culture, *Glia*, **8**, 71–76.
Pardridge, W.M., Triguero, D., Yang, J. and Cancilla., P.A. (1990) Comparison of *in vitro* and *in vivo* models of drug transcytosis through the blood-brain barrier, *J. Pharmacol. Expt. Therap.*, **253**, 884–891.
Pardridge, W.M., Yang, J., Eisenberg, J., and Mietus, L.J. (1986) Antibodies to blood-brain barrier bind selectively to brain capillary endothelial lateral membranes and to a 46K protein, *J. Cereb. Blood Flow Metab.*, **6**, 203–211.

Pardridge, W.M. (1991) Peptide drug delivery to the brain, Raven Press, New York, 1991, pp. 91–98.

Risau, W., Engelhardt, B., Wekerle, H. (1990) Immune function of the blood-brain barrier: incomplete presentation of protein (auto)antigens by rat brain microvascular endothelium *in vitro. J. Cell Biol.*, **110**, 1757–1766.

Roux, F., Durieu-Trautmann, O., Chaverot, N., Claire, M., Mailly, P., Bourre, J-M., Strosberg, A.D. and Couraud, P-O. (1994) Regulation of gamma-glutamyl transpeptidase and alkaline phosphatase activities in immortalized rat brain microvessel endothelial cells, *J. of Cell. Physiol.*, **159**, 101–113.

Rubin, L.L., Hall, D.E., Parter, S., Barbu, K., Cannon, C., Horner, H.C., Janatpour, M., Liaw, C.W., Manning, K., Morales, J., Tanner, L.I., Tomaselli, K.J. and Bard, F. (1991) A cell culture model of the Blood-Brain Barrier, *The Journal of Cell Biology*,, **115**, 1725–1735.

Tio, S., Deenen, M. and Marani, E. (1990) Astrocyte-mediated induction of alkaline phosphatase activity in human umbilical cord vein endothelium: An *in vitro* model. *Eur. J. Morphol.*, **28**, 289–300.

Townsend, L.E., Juleff, R.S., Bendick, P.J., Glover, J.L. (1991) Mitosis and angiogenesis in microwell endothelial cell culture, *In Vitro Cell. Dev. Biol.*, **27A**, 97–99.

van Bree, J.B.M.M., de Boer, A.G., Danhof, M., Ginsel, L.A. and Breimer, D.D. (1988) Characterization of an *in vitro* blood-brain barrier: effects of molecular size and lipophilicity on cerebrovascular endothelial transport rates of drugs, *J. Pharmacol. Exp. Ther.*, **247**, 1233–1239.

van Bree, J.B.M.M., de Boer, A.G., Danhof, M. and Breimer, D.D. (1992) Drug transport across the blood-brain barrier: II. Experimental techniques to study drug transport. *Pharm. Weekblad Sci. Ed.*, **14**, 338–348.

White, F.P., Dutton, G.R. and Norenberg, M.D. (1981) Microvessels isolated from rat brain: localization of astrocyte processes by immunohistochemical techniques, *J. Neurochem.*, **36**, 328–332.

II.3. RAT BRAIN MICROVESSEL ENDOTHELIAL CELLS: PRIMARY CULTURES AND IMMORTALIZED DIFFERENTIATED CELL LINE

D. LECHARDEUR, B. SCHWARTZ AND D. SCHERMAN

UMR 133 CNRS/Rhône-Poulenc Rorer, Centre de recherche de Vitry-Alfortville, Bâtiment Monod, 13, quai Jules Guesde, BP 14, 94403 Vitry/Seine, France

The use of enzymatic procedures for the isolation of rat brain (and bovine brain, see III.3) microvessels allows to obtain relatively pure primary cultures of endothelial cells. These cells are useful for biochemical and cellular analyses *in vitro*. They are also promising for obtaining a pharmacologically relevant *in vitro* model of the blood-brain barrier (BBB) which can be used for a preliminary selection of the brain penetration of drug candidates.

Brain microvessel endothelial cells might also be the starting material for obtaining an immortalized clonal cell line, which constitutes an attractive and user-friendly alternative *in vitro* model of the blood-brain barrier, provided that the cell line still expresses the differentiation markers of brain capillary endothelial cells. We have obtained a rat cell line immortalized by the SV40 $T\Delta t$ oncogene under the dependence of the human vimentin promoter, which possesses an endothelial phenotype and which expresses BBB characteristic markers when treated with the differentiating agent all-trans retinoic acid.

INTRODUCTION

In order to develop a convenient and versatile method to predict the brain penetration of drug candidates, major efforts are being spent by several groups to obtain an *in vitro* cellular model of blood-brain-barrier (BBB) (Rubin *et al.*, 1991; Dehouck *et al.*, 1992).

These models should display relevant *in vivo* characteristics:

I.   Establishment of intercellular tight junctions leading to a high transmonolayer electric resistance,
II. Lack of intracellular vesicules and of transcytotic traffic,
III. Expression of BBB specific enzymes or transporters such as gamma-glutamyl-transpeptidase (DeBault and Cancilla, 1980), the GLUT 1 glucose transporter (Boado *et al.*, 1990), or the *mdr* P-glycoprotein (Cordon-Cardo *et al.*, 1989, 1990).

In particular, the P-glycoprotein (Pgp) pumps drugs out of the multidrug resistant cells and thus prevents intracellular drug accumulation (Shimabuku *et al.*, 1992). Therefore, the polarized expression of Pgp on the apical side of the brain capillary endothelial cell monolayer should result in a vectorial basal-to-apical flux of Pgp substrates, such as that observed in the *in vitro* intestinal model made up of the Caco2 colon carcinoma cell line (Wils *et al.*, 1994, 1995). For BBB models, this should reflect the *in vivo* situation, in which the strictly apical Pgp expression in brain capillary endothelial cells is commonly thought to be responsible for an active brain to blood transport, through the BBB, of drugs such as vinca alkaloids, cyclosporin or other neurotoxic compounds, leading to a very low brain penetration (Greig *et al.*, 1990; Begley *et al.*, 1990). Relevant to this point, disruption of the mouse mdr1a P-glycoprotein has been shown to lead to an increased brain sensitivity to neurotoxic drugs (Schinkel *et al.*, 1994), indicating that the blood-brain barrier may indeed function as an active barrier against the brain penetration of many lipophilic compounds.

In order to study the expression of the *mdr* P-glycoprotein in primary cultures of endothelial cells of brain microvessels, and to develop immortalized endothelial cell lines, we have used enzymatic procedures for isolation of rat (and bovine, see III.3) brain capillary endothelial cells.

This chapter (and chapter III.3) aim to illustrate the techniques and strategies used to obtain proper primary cultures or clonal cell lines. For more specific cell biology techniques or for materials used, the reader should refer to Lechardeur and Scherman (1995), and Lechardeur *et al.* (1995).

PRIMARY CULTURE OF RAT BRAIN CAPILLARY ENDOTHELIAL CELLS

Rat cerebral endothelial cells were isolated from rat brains (OFA, Iffa Credo) according to a method modified from Hughes and Lantos (1986). In brief, animals were sacrificed under veterinary control and the brains collected. After the removal of surface vessels and meninges, cortical grey matter was minced and incubated in the preparation medium (DMEM 1 g/l glucose, 25 mM HEPES, 100 U/ml penicillin, 100 μg/ml streptomycin) containing 10 μg/ml DNAse 1 from bovine pancreas, and 1 mg/ml collagenase/dispase for 1 h at 37 °C.

After centrifugation at 1000 g for 10 min, the pellet was resuspended in a BSA solution (25% BSA final concentration in the preparation medium) and centrifuged at 2500 g for 10 min. Fat, cell debris and myelin floating on BSA were discarded and the pellet containing microvessels was resuspended in the collagenase/dispase as described above for 1 to 3 h at 37 °C in a shaking waterbath. Microvessels were then separated from cell debris and erythrocytes by centrifugation on a pre-established 50% Percoll gradient at 2000 g for 10 min. Capillaries were then washed with the preparation medium and resuspended in culture medium consisting of 45% αMEM with ribonucleosides and desoxyribonucleosides, 45% Nutrient Mixture F-10(HAM) supplemented with 10% foetal calf serum or a mixture of 5% foetal calf serum and 5% rat serum, 100 μg/ml streptomycin, 100 U/ml penicillin, 2 mM L-glutamine and 5 ng/ml bFGF.

Endothelial cells were seeded onto plastic dishes which had been pre-coated overnight with rat tail collagen (250 μg/ml in 0.1 N acetic acid), rinsed with PBS, then coated with fibronectin from bovine plasma (25 μg/ml) or with extracellular matrix produced by bovine corneal endothelial cells (Gospodarowicz *et al.*, 1984).

The morphology of rat brain capillary endothelial cells in primary culture was similar to that of bovine brain (Figure 1). These cells expressed the marker of endothelial cells Von Willebrand factor (factor VIII-related antigen) (Figure 2). They were frequently contaminated by cells of different morphology, presumably astrocytes, pericytes, or smooth muscle cells (Figure 1).

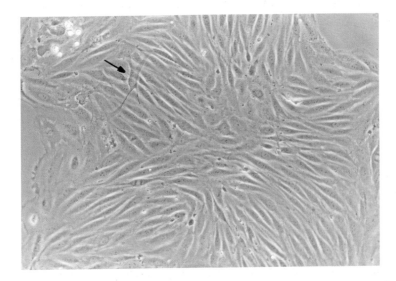

Figure 1 Rat brain capillary endothelial cells in primary culture. Spindle-shaped rat endothelial cells were obtained as described; they were sometimes contaminated by cells which developed over the endothelial cell monolayer (arrow), and which became predominant after the second passage. Phase contrast microscopy (×100).

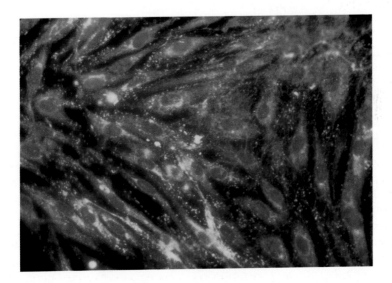

Figure 2 Expression of the Von Willebrand factor (factor VIII related antigen) in a primary culture of rat brain capillary endothelial cells. Indirect immunofluorescence with rabbit anti-human Von Willebrand factor antibody. Labelling is observed in a granular population concentrated in the perinuclear zone.

Results Obtained with Primary Cultures

Several points in the protocol presented here deserve some comments on the methodologies.

First, concerning the centrifugation step in 25% BSA following the first collagenase-dispase digestion: we cannot ascertain that 25% BSA, which is fairly high and intended to eliminate cell debris and myelin, leads to more purified capillaries. We prepared, for biochemical studies, several bovine capillary samples in 25% BSA, after mechanical disruption and without collagenase/dispase. We then observed a 20–30 purification yield. We prepared once bovine capillaries on 18% BSA. After purification, the purity seemed to be slightly lower (enrichment yield of about 10). However, the influence of the BSA concentration on the purity of the obtained material was not systematically investigated.

Second, we observed that the coating of plastic dishes for seeding isolated capillaries was critical. For rat cells the best results were obtained with the extracellular matrix produced by bovine corneal endothelial cells. Alternatively, rat tail collagen can be used; the two techniques give equivalent results (250 μg/ml in 0.1 N acetic acid, 250 μl/cm^2, for 2 h at 37 °C, or 3 mg/ml, 100 μl/cm^2 neutralized under NH_3 vapours). Unsatisfactory results were obtained with air-dried collagen.

Third, isolation of microvessels might be reached by an alternative procedure. Briefly, the brain tissue is homogenized with a Potter-Elvejhem homogenizer (glass/glass, 75–150 μm, 10 up and down strokes in DMEM). The sample is centrifuged (2000 g, 10 min), the pellet is mixed with Dextran 70,000 (17% final concentration) and centrifuged again at 3000 g for 15 min. The pellet is resuspended in HBSS with 1% BSA, then filtered through a 150 μm nylon sieve. The filtrate is then layered on a column (diameter: 1 cm; height: 2 cm) of glass beads (diameter: 200–300 μm). The column is washed with Ca/Mg free HBSS, and all effluents discarded. The glass beads are then poured into a large volume of Ca/Mg-free HBSS, and the capillaries are released from the beads upon gentle shaking, collected by centrifugation and subjected to the second collagenase/dispase digestion, as described above.

The two first protocols for rat (and bovine, see III.3) brain endothelial cell primary cultures have been used in the following experiments:

- to study the expression of the P-glycoprotein in primary cultures of bovine and rat brain endothelial cells by immunoblots (after scrapping confluent monolayers and preparing membrane fractions), and by functional experiments (incorporation and efflux of Pgp substrates such as vinblastin) (Lechardeur and Scherman, 1995);
- immortalization of rat brain endothelial cells by transfection with SV40 T gene (see below; Lechardeur et al., 1995).

Establishment of an Immortalized Endothelial Cell Line from Rat Brain Capillary

Cell microinjection and the obtainment of the CR3 cell line

Rat brain capillary endothelial cells in primary culture were microinjected with a linear DNA construct containing the thermosensitive large T gene tsA 58 of SV40 under the control of a 971 base pair deletion mutant of the human vimentin gene promoter (Pinçon-Raymond *et al.*, 1991; Schwartz *et al.*, 1991).

Five hundred to 2000 nuclei of cells selected for a typical capillary endothelial cell morphology were individually microinjected with a 10 ng/ml DNA solution in Tris-EDTA buffer (10 mM Tris, 1 mM EDTA, pH 7.2) (Lechardeur *et al.*, 1995). The volume injected was restricted to about a two-fold increase in the size of the nucleus. The cells were then washed with the medium, grown in DMEM containing 10% FCS, and observed for three weeks. Whereas the majority of the culture underwent senescence, colonies of actively dividing cells showing an endothelial morphology were picked up by local trypsinization, and expanded. Cell lines were established after cloning twice by limiting dilution, and were then characterized. Cells were grown in DMEM, 4.5 g/l glucose, 2 mM glutamine, 100 μg/ml streptomycin, 100 U/ml penicillin and 10% fetal calf serum, in humidified 5% CO_2/95% air at 37 °C.

After screening of about 20 clones for endothelial cell characteristics (Von Willebrand factor expression and morphology), one cell line, named CR3, was selected for further investigation. As shown in Figure 3, more than 95% of the transformed cells at passage 6 synthetized the nuclear large T protein, suggesting that the microinjected DNA sequence had been integrated in the genome of the cells. The CR3 cell line expressed a dense network of vimentin filaments, was able to grow at clonal density, was passaged more than 50 times, had a division time of about 24 h in 10% or 5% FCS. A sub-population could be obtained, which grew in less than 1% FCS, but after a long selection from the parental line. At the saturation density, the cells detached from the support very rapidly, less than two days after reaching confluency. Stabilization of the confluent monolayers was attempted by maintaining the cells at 39 °C in order to inhibit the thermosensitive large T protein. This treatment did not dramatically decrease cell growth and had no effect on the stability of the monolayers at confluency. The CR3 cells were not able to grow in semi-solid media. This cell line showed an abnormal number of chromosomes.

Differentiated endothelial characteristics of the CR3 cell line

CR3 cells formed a monolayer of cells characterized by an elongated spindle-shaped morphology, similar to that of primary endothelial cells from capillary origin (Figure 1 in chapter III.3; Figures 1 and 4 in this chapter). The endothelial marker factor VIII-related antigen (the Von Willebrand factor) was localized by immuno-cytofluorescence at the perinuclear level, as commonly observed in endothelial cells (Dorovini-Zis and Huynh, 1992). The expression of the Von Willebrand factor in the CR3 cell line seemed to be maintained for more than 30 passages. The CR3 cells do not express the α-smooth muscle cell actin, a classical marker of pericytes and smooth muscle cells (Verbeek *et al.*, 1994), cell types which frequently contaminate primary cultures of cerebral endothelial cells (Abbott *et al.*, 1992). The CR3 cell line also took up acetylated LDL, a property characteristic of endothelial cells and

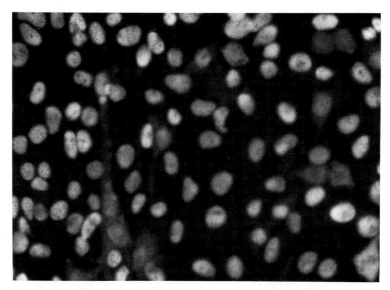

Figure 3 Expression of the large T antigen in the CR3 cell line. Immunostaining of the CR3 cell line was performed using a mouse monoclonal antibody against the SV40 large T antigen; nearly all cells show a clearly marked nuclear fluorescence.

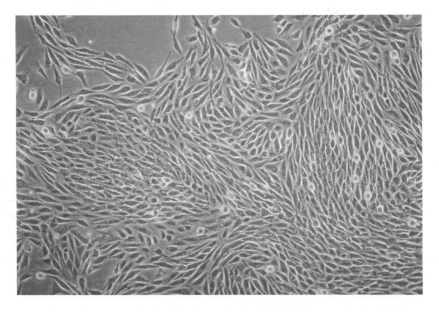

Figure 4 CR3 cell line endothelial cells. Phase contrast microscopy, showing a typical spindle shape of the cells, reminiscent of brain microvessel endothelial cells in primary culture.

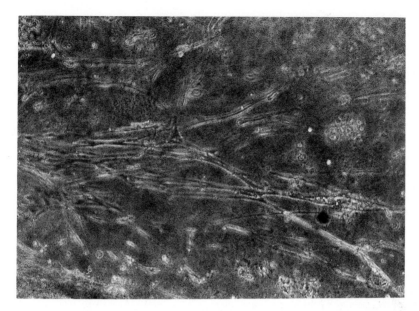

Figure 5 Phase contrast microscopy of CR3 cells seeded in a three dimensional collagen matrix. After 3 to 4 weeks, cells develop particular elongated structures reminiscent of vessel formation.

macrophages (Voyta *et al.*, 1984), and expressed the angiotensin-I converting enzyme mRNA, another endothelial marker. When seeded in a three dimensional collagen matrix, these cells develop, after three to four weeks, elongated tubular formation, suggesting an ability to reconstitute capillary-like structures (Figure 5).

Blood-brain barrier differentiation markers of the CR3 cell line

The γ-glutamyl transpeptidase enzyme activity is a classical marker of the blood-brain barrier endothelial cells (Orlowski *et al.*, 1974). In the CR3 cells, the activity of the γ-glutamyl transpeptidase appeared to be expressed at low levels in comparison with rat brain capillaries and with rat brain cortex. However, when the cells at confluency were subjected for 24 h to 5 μM all-*trans*-retinoic acid, which is a chemical cell differentiating agent (Leid *et al.*, 1992), the γ-glutamyl transpeptidase activity was increased. To verify that the enhancing effect of this chemical inducer was specific of the CR3 cells, the same experiment was assayed on the ECV 304 human umbilical vein endothelial cell line: no change in γ-glutamyl transpeptidase activity was detected on the ECV 304 human umbilical vein endothelial cell line, indicating the specificity of the all-*trans*-retinoic acid induction.

No detectable P-gp expression was found in the CR3 cells grown in the classical culture medium. On the contrary, when cells were subjected to all-*trans*-retinoic acid, P-gp expression was enhanced, and the P-glycoprotein functional activity was monitored by assaying the efflux pump activity of this protein on [^{3}H]vinblastine, a vinca alkaloid anticancer drug known to be a substrate of the P-gp (Pearce *et al.*, 1989).

Retinoic acid treated CR3 cells might thus represent a useful tool for biological and pharmacological research related to the blood-brain barrier (Lechardeur *et al.*, 1995).

CONCLUDING REMARKS

Our experience has shown that rat cells are not so easy to manipulate as bovine cells with which we have obtained the largest quantity of cells, in cultures that divided more quickly and with less numerous contaminating cells. (see III.3).

For the rat endothelial cell cultures, the best results have been obtained using rat serum (prepared in the laboratory) and corneal endothelial cell extracellular matrix (as described above).

The use of primary cultures seems to be limited by the presence of contaminating cells which did not allow us to obtain tight monolayers in dual chamber systems. This prompted us and others (Durieu-Trautmann *et al.*, 1991) to develop clonal immortalized cell lines, which might present several advantages:

- a cell line is user-friendly: it avoids the tedious isolation of brain capillaries for obtaining primary cultures,
- homogeneity and reproducibility of results (allowing comparison of the results obtained by several laboratories),
- a clonal cell line might be manipulated to express proper BBB characteristics, for instance by using differentiation agents such as retinoic acid, or by introducing BBB-specific genes associated to selection markers such as the neomycin resistance gene.

The establishment of the CR3 cell line might prove to be a more useful tool for the study of *in vitro* brain penetration (which could be performed on the same species i.e. the rat) than the method used for *in vivo* preclinical experiments. However, an unsolved problem remains, since tight monolayers displaying a high electrical resistance could not be obtained with CR3 cells. To our knowledge, such a differentiation level (i.e. establishment of tight junctions) has not been reported for any other immortalized BBB endothelial cell lines, although the effect of differentiating agents has not been studied systematically.

An alternative to *in vitro* immortalized cell lines could be cell lines derived from transgenic animals bearing immortalizing genes. This approach has been shown to generate cell lines that are difficult to produce *in vitro*, and which generally display a more differentiated phenotype (Jat, 1986; McKay *et al.*, 1988; Paul *et al.*, 1988; Vicart *et al.*, 1994).

ACKNOWLEDGEMENTS

This work was sponsored by the CNRS, the Rhône Poulenc Rorer Society, and the BIOAVENIR program of French Ministry of Education and Research. We are indebted to Dr. Monique Santarromana for excellent help and suggestions in the realization of this commentary.

REFERENCES

Abbott, N.J., Hughes, C.C.W., Revest, P.A. and Greenwood, J. (1992) Development and characterisation of a rat brain capillary endothelial culture: Towards an *in vitro* blood-brain barrier. *J. Cell Sci.*, **103**, 23–37.

Audus, K.L. and Borchardt, R.T. (1986) Bovine brain microvessels endothelial cell monolayers as a model system for the blood-brain barrier. *Ann. N. Y. Acad. Sci.*, **507**, 9–18.

Begley, D.J., Squires, L.K., Zlokovic, B.V., Mitrovic, D.M., Hughes, C.C.W., Revest, P.A. and Greenwood, J. (1990) Permeability of the blood-brain barrier to the immunosuppressive cyclic peptide cyclosporin A. *J. Neurochem.*, **55**, 1222–1230.

Boado, R.J. and Pardridge, W.M. (1990) The brain-type glucose transporter mRNA is specifically expressed at the blood-brain barrier. *Biochem. Biophys. Res. Commun.*, **166**, 174–179.

Cordon-Cardo, C., O'Brien J.P., Casals, D., Rittman-Grauer, L., Biedler, J.L., Melamed, M.R. and Bertino, J.R. (1989) Multidrug-resistance gene (P-glycoprotein) is expressed by endothelial cells at blood-brain barrier sites. *Proc. Natl. Acad. Sci.*, **86**, 695–698.

Cordon-Cardo, C., O'Brien, J.P., Boccia, J., Cassals, D., Bertino, J.R., and Melamed, M.R. (1990) Expression of the multidrug resistance gene product (P-glycoprotein) in human normal and tumor tissues. *J. Histochem. Cytochem.*, **38**, 1277–1287.

DeBault, L.E. and Cancilla, P.A. (1980) Gamma-glutamyl transpeptidase in isolated brain endothelial cells: Induction by glial cells *in vitro*. *Science*, **207**, 653–655.

Dehouck, M.P., Méresse, S., Dehouck, B., Fruchart, J.C. and Cecchelli, R. (1992) *In vitro* reconstituted blood-brain barrier. *J. Controlled Release*, **21**, 81–92.

Dorovini-Zis, K. and Huynh, H.K. (1992) Ultrastructural localization of factor VIII-related antigen in cultured human brain microvessel endothelial cells. *J. Histochem. Cytochem.*, **40**, 689–696.

Durieu-Trautmann, O., Foignant-Chaverot, N., Perdomo, J., Gounon, P., Strosberg, A.D. and Couraud, P.O. (1991) Immortalization of brain capillary endothelial cells with maintenance of structural characteristics of the brain barrier. *In Vitro Cell Dev. Biol.*, **27A**, 771–778.

Gospodarowicz, D., Lepine, J., Massoglia, S. and Wood, I. (1984) Comparison of the ability of basement membranes produced by corneal endothelial and mouse-derived endodermal PF-HR-9 cells to support the proliferation and differentiation of bovine kidney tubule epithelial cells *in vitro*. *J. Cell. Biol.*, **99**, 947–961.

Greig, N.H., Soncrant, T.T., Shetty, U., Momma, S., Smith, Q.R. and Rappoport, S.I. (1990) Brain uptake and anticancer activities of vincristine and vinblastine are restricted by their low cerebrovascular permeability and binding to plasma constituents in rat. *Cancer Chemother. Pharmacol.*, **26**, 263–268.

Hughes, C.C.W. and Lantos P.L. (1986) Brain capillary endothelial cells *in vitro* lack surface IgG Fc receptors. *Neurosci. Lett.*, **68**, 100–106.

Jat, P., Cepko, C., Mulligan, R. and Sharp, P. (1986) Recombinant retroviruses encoding simian virus 40 large T antigen and polyomavirus efficiently establish primary fibroblasts. *J. Virol.* **59**, 746–750.

Lechardeur, D., Schwartz, B., Paulin, D. and Scherman, D. (1995) Induction of blood-brain barrier differentiation in a rat brain derived endothelial cell line. *Exp. Cell Res.*, **220**, 161–170.

Lechardeur, D. and Scherman, D. (1995) Functional expression of the P-glycoprotein *mdr* in primary cultures of bovine capillary endothelial cells. *Cell Biol. Tox.*, **11**, 219–230.

Leid, M., Kastner, P. and Chambon, P. (1992) Multiplicity generates diversity in the retinoic acid signalling pathways. *Trends Biochem. Sci.*, **17**, 427–432.

McKay, K., Striker, L., Elliot, S., Sprinker, C., Brinster, R. and Striker, G. (1988) Glomerular epithelial, mesangial and endothelial cell lines from transgenic mice. *Kidney Int.*, **33**, 677–684.

Orlowski, M., Sessa, G. and Green, J.P. (1974) Gamma-glutamyl transpeptidase in brain capillaries: Possible site of a blood-brain barrier for amino-acids. *Science*, **184**, 66–68.

Paul, D., Höhne M., Pinkert, C., Piasecki, A, Ummelman, E. and Brinster, R. (1988) Immortalized differentiated hepatocyte lines derived from transgenic mice harboring SV40 T-antigen genes. *Exp. Cell. Res.*, **175**, 354–362.

Pearce, H.L., Safa, A.R., Bach, N.J., Winter, M.A., Cirtai, M.C. and Beck, W.T. (1989) Essential features of the P-glycoprotein pharmacophore as defined by a series of reserpine analogs that modulate multidrug resistance. *Proc. Natl. Acad. Sci. USA*, **86**, 5128–5132.

Pinçon-Raymond, M., Vicart, P., Bois, P., Chassande, O., Romey, G., Varadi, G., Li, Z.L., Ladzunski, M., Rieger, F. and Paulin, D. (1991) Conditional immortalization of normal and dysgenic mouse muscle cells by the SV40 large T antigen under the promoter vimentin control. *Dev. Biol.*, **148**, 517–528.

Rubin, L.L., Hall, D.E., Porter, S., Barbu, K., Cannon, C., Horner, H.C., Janatpour, M., Liaw, C.W., Manning, K., Morales, J., Tanner, L.I., Tomaselli, K.J. and Bard, F. (1991) A cell culture model of the blood-brain barrier. *J. Cell. Biol.*, **115**, 1725–1735.

Schinkel, A.H., Smit, J.J., Van Tellingen, O., Beijnen, J.H., Wagenaar, E., Van Deemter, L., Mol, C.A., Nan der Valk, M.A., Robanus-Maandag, E.C., Te Riele, H.P., Berns, A.J. and Borst, P. (1994) Disruption of the mouse mdr1a P-glycoprotein gene leads to a deficiency in the blood-brain barrier and to increased sensitivity to drugs. *Cell*, **77**, 491–502.

Schwartz, B., Vicart, P., Delouis, C. and Paulin, D. (1991) Mammalian cell lines can be efficiently established *in vitro* upon expression of the SV40 large T antigen driven by a promoter sequence derived from the human vimentin gene. *Biol. Cell*, **73**, 7–14.

Shimabuku, A.M., Nishimoto, T., Ueda, K. and Komano, T. (1992) P-glycoprotein. *J. Biol. Chem.*, **267**, 4308–4311.

Verbeek, M.M., Otte-Hölle, I., Wesseling, P., Ruiter, D.J. and De Waal, R.M. (1994) Induction of smooth muscle actin expression in cultured human brain pericytes by transforming growth factor-$\beta1$. *Am. J. Pathol.*, **144**, 372–382.

Vicart, P., Schwartz, B., Vandewalle, A., Bens, M., Delouis, C., Panthier, J.J., Pournin, S., Babinet, C. and Paulin, D. (1994) Immortalization of multiple cell types from transgenic mice using a transgene the vimentin promoter and a conditional oncogene. *Exp. Cell Res.*, **214**, 30–45.

Voyta, J.C., Via, D.P., Butterfield, C.E. and Zetter, B.R. (1984) Identification and isolation of endothelial cells based on their increased uptake of acetylated low density lipoprotein. *J. Cell Biol.*, **99**, 2034–2040.

Wils, P., Phung-Ba, V., Warnery, A., Lechardeur, D., Raeissi, S., Hidalgo, I.J. and Scherman, D. (1994) Polarized transport of docetaxel and vinblastine mediated by P-glycoprotein in human intestinal epithelial cell monolayer. *Biochem. Pharmacol.*, **48**, 1528–1530.

Wils, P., Phung-Ba, V., Warnery, A., Raeissi, S., Hidalgo, I. and Scherman, D. (1995) Interaction of pristinamycin IA with P-glycoprotein in human intestinal epithelial cells. *Eur. J. Pharmacol.*, **288**, 187–192.

II.4. DRUG METABOLIZING ENZYME ACTIVITIES IN AN *IN VITRO* MODEL OF RAT BLOOD-BRAIN BARRIER

A. MINN,[1] D. GRADINARU,[2] G.F. SULEMAN,[1]
M. CHAT,[1] C. BAYOL-DENIZOT[1] AND P. LAGRANGE[1]

[1] *CNRS URA 597-30 rue Lionnois, 54000 Nancy, France*
[2] *National Institute of Gerontology and Geriatrics, "Ana Aslan" – 9,*
Str Manastirea Caldarusani, 78178 Bucharest, Romania

The blood-brain barrier (BBB) possesses drug-metabolizing enzyme activities which provide a complementary protection to the brain, as a xenobiotic must first penetrate the BBB before its neurological effects become apparent. The biotransformation of these molecules inactivates some potentially toxic molecules before their entry into the central nervous system (CNS), and may also prevent their influx by increasing their polarity. The presence of this metabolic BBB has been demonstrated using isolated rat brain microvessels, which contain pericytes and astrocytic feet fragments. In this chapter we describe the use of primary cultures of rat cerebrovascular endothelial cells for a further characterization of the specific metabolic activities of these cells toward drugs that are model substrates of several enzymatic systems. Primary cultures of cerebrovascular endothelial cells display high activities of some of these enzymes, thus confirming the presence of an *in vivo* metabolic blood-brain barrier.

INTRODUCTION

The brain possesses unique and efficient protective systems controlling the passage of drugs and chemicals from the blood into the cerebral tissue, due to the presence of tight junctions between endothelial cells as well as the absence of transcytosis vesicles (Bates, 1985; Joó, 1985; Oldendorf, 1977). Brain capillaries are not fundamentally porous, thus the passage of chemicals from the blood into the brain occurs through the plasma membranes and cytosol of endothelial cells. Two categories of protective systems may alter the exchanges of molecules between the cerebral circulation and the CNS.

The first is a physical barrier, mainly related to the lipophilic characteristics of the permeable drug. As the cerebral microvasculature exhibits complex tightly closed junctions between endothelial cells, at least three membrane systems (luminal and antiluminal plasma membranes from endothelial cells, basement membrane, and also probably astrocyte feet membrane) divide the neuronal space from the systemic circulation by their proteolipidic properties. In fact, the entry of polar molecules into the brain is prevented or retarded, as these membranes are highly lipophilic, but nutrients like glucose, carboxylic acids and several amino acids possess specific facilitated or active blood-brain transport mechanisms (Joó, 1985; Pardridge, 1983). Consequently, nonpolar, lipid-soluble molecules can easily penetrate the brain, a property enabling the design of specific neurotropic drugs. However, this property also allows the access of some toxins to the brain.

The second is a metabolic barrier, formed by the enzymes of the endothelial cytosol which metabolize some of the permeable molecules recognized as substrates.

The presence of dopa decarboxylase, monoamine oxidase-A and -B, transaminases, alkaline phosphatase and adenosine deaminase activities in brain capillaries prevents the entry of circulating molecules with potential or established neuroactivity, like neurotransmitters, neuromodulators, or their proximate precursors, that could interfere with the normal functioning of the brain (Joó, 1985; Oldendorf, 1977; Pardridge, 1983). Moreover, recent data indicate that isolated brain capillaries contain significantly higher activities of enzymes involved in drug metabolism than the brain parenchyma itself. Cytochrome P450 (CYP) levels are practically identical (Ghersi-Egea *et al.*, 1988a), but both isoforms CYP1A and CYP2B activities, which have been shown to metabolize polycyclic aromatic hydrocarbons as well as drugs like caffeine and theophylline (Campbell *et al.*, 1987), imipramine (Lemoine *et al.*, 1993) and cocaine (Pasanen *et al.*, 1995) in the liver, display higher activities than in brain parenchyma (Perrin *et al.*, 1990). A greater formation of morphine from codeine has been reported in a microvessel fraction than in microsomes prepared from the rat brain, suggesting that the efficient metabolic capacity of CYP2D of endothelial cells may have in some instances a prodrug-activating function (Chen *et al.*, 1990). The possible implication of CYP2E1, the isoform involved in ethanol and ketone metabolism, in the vascular biotransformation of arachidonic acid to vasoactive compounds has also been suggested (Hannsson *et al.*, 1990), thus proposing a physiological role for this isoform. Moreover, other drug metabolizing enzyme activities, like NADPH-cytochrome P450 reductase, 1-naphthol glucuronosyl-transferase (UGT) and microsomal epoxide hydrolase, are respectively 2, 15 and 5 times higher than in the brain (Ghersi-Egea *et al.*, 1988a, 1994; Suleman *et al.*, 1993). These enzymes should therefore protect the brain from direct chemical insult. They may also modify drugs during their transport to the brain, resulting in an inactivation of their pharmacological properties, but also possibly in the formation of neurotoxic side products. These activities could also be useful therapeutic tools for the activation of neuroactive prodrugs. And lastly, cerebrovascular endothelial cells are the only cell groups in the CNS that are continuously exposed to blood constituents, including drugs, but also endogenous regulatory factors, like cytokines and growth factors, which may modify BBB function via surface receptors on their luminal membrane (Kanda *et al.*, 1995).

As isolated brain microvessel fraction usually contains numerous contaminating elements, like mid-size blood vessels, pericytes and adhering astrocyte fragments, the use of primary cultures of cerebrovascular endothelial cells allows specific studies of the drug metabolizing properties of the monocellular layer forming a convenient *in vitro* model of the blood-brain barrier.

MATERIALS AND METHODS FOR SYSTEM ESTABLISHMENT

Animals

Ten adult male Sprague-Dawley rats, weighing 200–230 g, obtained from Iffa-Credo, St Germain-sur-l'Arbresle, France, were used for each experiment. The animals were acclimatized to laboratory conditions (normal 12 h/12 h light cycle, temperature $22 \pm 1\,°C$) for one week before the start of experiments, and received specific rat chow and tap water *ad libitum*. Food was removed 12 h before the experiment, in order to decrease blood lipid concentration.

Each rat was killed by decapitation, and the head was immersed in ethanol before removing the brain from skull. Tools were immersed in ethanol before each brain removal, and the experimenters wore sterile surgical gloves. The brains were dropped into DMEM/F12 1/1 mixture containing 100 IU/ml penicillin and 100 mg/ml streptomycin. Blood was collected in a tube containing 0.5 ml sodium citrate 0.38% and centrifuged at 3000 g for 5 min to sediment erythrocytes. The supernatant was then centrifuged at 12,500 g for 20 min to discard blood platelets. The serum was then coagulated by the addition of 1 volume of $CaCl_2$ (200 mM) in NaCl (0.9%) to 9 volumes of serum at 37 °C, and centrifuged at 12,500 g for 30 min at 4 °C. The supernatant was dialysed 3 times for 8 h against PBS buffer, decomplemented by heating at 56 °C for 30 min and sterilized by ultrafiltration (Micropore, 0.22 μm) before freezing for further use.

Primary Cultures of Rat Cerebrovascular Endothelial Cells

The pial membranes and large blood vessels on the brain surface were carefully removed, and the cortices were dissected and homogenized with a hand glass homogeneizer (Dounce) in 30 ml DMEM Mix F12 containing antibiotics. The homogenate was then centrifuged for 5 min at 600 g. The sediment was removed and dispersed in 30 ml of collagenase/dispase solution (1 mg/ml) prepared in the same medium. The suspension was then incubated for 1 h at 37 °C under slow agitation. After enzymatic treatment, the suspension was again centrifuged for 5 min at 1000 g. The supernatant was discarded, and the sediment resuspended in 50 ml HBSS. This suspension was divided in two parts, and each part was completed up to 50 ml with a solution of 26% dextran and centrifuged for 15 min at 2500 g to sediment capillaries. The sediment was washed with HBSS and recentrifuged at 1000 g.

To dissociate endothelial cells, isolated capillaries were suspended in 20 ml of the collagenase/dispase solution and incubated for 3 h at 37 °C under slow agitation. Meanwhile, a Percoll gradient was prepared as follows: 12 ml of HBSS (2×) and 12 ml Percoll were poured in sterile tubes, and centrifuged for 1 h at 18,000 g and 4 °C in order to form the concentration gradient. The tubes were left at 4 °C until the end of capillary enzymatic dissociation. The suspension was centrifuged at 1000 g for 5 min, the sediment resuspended in 2 ml HBSS and layered onto the top of the Percoll gradient. The gradient was centrifuged at 2000 g for 15 min, and the endothelial cell fraction was carefully removed by aspiration from the middle of the gradient and centrifuged at 600 g for 5 min. The seeding was performed on collagen (250 μg/ml) — coated six-well plates, with a density of one rat brain endothelial cell in DMEM/F12 containing antibiotics, 5% fetal bovine serum and 5% rat serum, for 10 cm^2. Culture medium was fully replaced twice a week, and Figure 1A shows cultured rat cerebrovascular cells close to confluence.

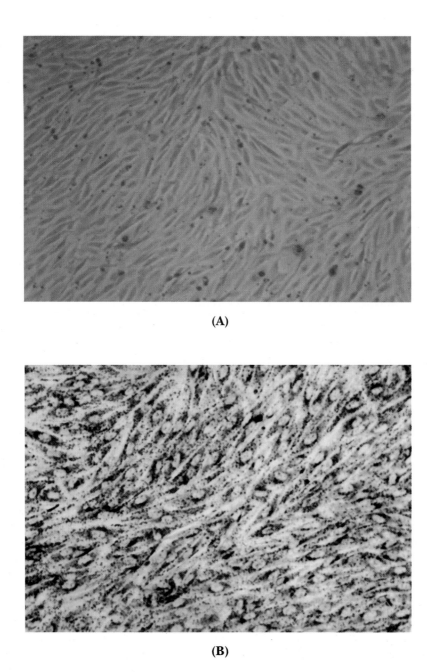

(A)

(B)

Figure 1 Micrographs of confluent (15-day-old) rat cerebrovascular endothelial cells. (A) without staining, × 200, (B) after Factor VIII-immunostaining, × 400.

Characterization of Endothelial Cells

After fixation by methanol at $-20\,°C$ for 20 min, the cells were washed with PBS and incubated overnight with anti-Factor VIII antibody (Sigma, diluted 1/1000 in PBS). Rabbit anti-IgG alkaline phosphatase (Sigma, diluted 1/100 in PBS) was used as secondary antibody, and was incubated for 3 h and revealed with BCIP/NBT. This immunocharacterization showed that 80–90% of confluent monocellular layers were Factor VIII-positive cells (Figure 1B).

Measurement of Drug-metabolizing Enzyme Activities

Most of these activities are localized in the endoplasmic reticulum of cells, therefore their activities are usually measured on the microsomal fraction, that is the subcellular fraction containing the cytoplasmic membranar systems (Perrin et al., 1990). Nevertheless, as the quantity of proteic material obtained from cell cultures is rather low, the activities have been measured directly on cell homogenates. Confluent cultures were obtained usually at 11 days, the cell layer was harvested from the culture dish and suspended in a volume of 100 µl of 0.32 M sucrose, 1 mM EDTA, 10 mM Tris-HCl buffer, pH 7.4. The cell suspension was carefully homogenized using a glass Dounce hand homogeneizer by several up and down movements of the pistle. After measurement of the protein content, aliquots of the homogenates were used as samples for enzymatic activity measurements.

Cytochromes P450-dependent activities

Cytochromes P450 are multiple-function monooxygenases able to catalyse C-, N-, S-oxygenations, as well as dealkylation, epoxydation or peroxidation reactions. These activities lead generally to the formation of less toxic but more hydrophilic products. They may also form toxic or reactive metabolites such as epoxides and nitrosamines as well as activated forms of oxygen, especially superoxide anions and hydroxyl radicals. Moreover, cytochromes P450 are involved in the metabolism of endogenous molecules, like eicosanoids (Hannsson et al., 1990; Smith, 1989) and neurosteroids (Hu et al., 1987; Le Gascogne et al., 1989; Walther et al., 1987), which possess important physiological functions, related to cerebral bood flow regulation, modulation of GABA receptors activity (Majewska and Schwartz, 1987), and control of reproductive behaviour.

CYP1A and CYP2B activities are mainly involved in the metabolism of exogenous molecules like aryl hydrocarbons (Perrin et al., 1990), drugs like imipramine (Lemoine et al., 1993) and also probably some food additives, eventually with neurotropic activity, like caffeine, theophylline or nicotine, which are partially metabolized in both rat and human liver by CYP1A. These activities were measured using model substrates, 7-ethoxyresorufin for CYP1A and 7-benzoxyresorufin for both CYP1A and CYP2B (Perrin et al., 1990). 20 to 50 µg of homogenate protein were poured in a cuvette containing 1 ml of 15 mM $MgCl_2$, 0.1 mM EDTA, 50 mM HEPES buffer, pH 7.5. The formation of resorufin by NADPH-dependent CYP activities was followed by spectrofluorimetric measurement, exactly as described in (Perrin *et al.*, 1990).

NADPH-cytochrome P450 reductase

This flavoprotein supplies two electrons from NADPH to cytochrome P450, allowing the normal functioning of the monooxygenase. Nevertheless, this reductase displays relatively high activities in the brain (Ghersi-Egea *et al.*, 1989; Haglund *et al.*, 1984) and in isolated cerebral microvessels (Ghersi-Egea *et al.*, 1988a, 1994) and catalyses the monoelectronic reduction of some drugs (Lagrange *et al.*, 1994; Pasanen *et al.*, 1995). In normoxic conditions, this may therefore result in the production of toxic oxygen radicals (Bayol-Denizot *et al.*, 1996; Lagrange *et al.*, 1994). The NADPH-cytochrome P450 reductase activity was measured at 25 °C and $\lambda = 450$ nm, with 10 μM cytochrome c as the substrate, in a 330 mM K-phosphate, 0.1 mM K-EDTA, pH 7.40 buffer as described previously (Ghersi-Egea *et al.*, 1989). The reaction was started by the addition of NADPH at a final concentration of 46 μM. A possible interfering non-enzymatic reduction of the cytochrome c was prevented by the use of a double-beam spectrophotometer with a control cuvette containing all reagents except NADPH.

UDP-glucuronyltransferases (UGT)

This family of enzymes catalyzes the conjugation of a glucose derivative, uridyl-diphospho (UDP)-glucuronic acid, to a series of substrates, among which are hydroxylated products resulting from CYP activities (Dutton, 1980). UGTs are intrinsic membrane proteins localized in the endoplasmic reticulum (ER) of the cells, the active site being located within the ER lumen. Therefore, UGTs activities are highly latent and can be increased by non-ionic detergent treatment. In the rat brain capillaries, only the isoform UGT1*6, catalyzing the conjugation of 1-naphthol and 4-methylumbelliferone as model substrates, has been detected (Ghersi-Egea, 1988a, 1994). Nevertheless, the possible involvement of some other UGT isoforms in the synthesis of complex glycolipids expressed in cerebrovascular endothelial cells remains to be demonstrated (Kanda *et al.*, 1995). The activities measured in cerebrovascular endothelial cells are as high as those measured in primary cultures of rat cortex neurones (Gradinaru *et al.*, 1995).

Cytosolic sulphotransferase

The formation of sulphate conjugates is a common biochemical reaction well recognized as an important feature of chemical defence. Sulphotransferases catalyse the transfer of a sulfonate moiety from the donor molecule 3′-phosphoadenosine-5′-phosphosulphate (PAPS) to a wide variety of phenolic compounds and represent quantitatively, with glucuronidation, the two major conjugation processes. Sulphate conjugation being a high affinity system as compared to glucuronidation, would conceivably be more significant in the inactivation of endogenous neurotransmitters amines, hormones and some drugs (Karoum *et al.*, 1993). Most studies on brain sulphotransferase activity have been concerned with the conjugation of biogenic amines and their metabolites and it has been shown that sulphation of some endogenous catecholamine metabolites facilitates their efflux from the brain by a probenicid-sensitive transport mechanism (Meek and Neef, 1972). Moreover, some sulpho-conjugates of steroids are active on central receptors (Majewska *et al.*, 1987). Nevertheless, a xenobiotic sulphate conjugation has been demonstrated only in primary cultures of bovine brain microvessel endothelium (Baranczyk-Kuzma *et al.*, 1989). This provides a strong argument for the physiological role of the metabolic BBB in regulating drug movements between the blood and the CNS.

SPECIFIC APPLICATIONS

Cultured cerebrovascular endothelial cells retain most of the structural properties they display *in vivo*. The formation of tight junctions between cells at confluence allows the formation of a monocellular layer with high electrical resistivity, providing a convenient model of the physical blood-brain barrier *in vitro*. The cultured cells also possess other functional properties, as scavenger receptors (de Vries *et al.*, 1993; Joó, 1985). As we developed recently the concept of a metabolic blood-brain barrier (Ghersi-Egea *et al.*, 1988a,b, 1994; Minn *et al.*, 1991, 1993) able to protect — at least in part — the brain against both neuroactive molecules and potentially neurotoxic chemicals, our purpose was the characterization of the enzymatic systems able to metabolize lipophilic molecules to more polar and inactive products. These hydroxylated or conjugated products should be less permeable to the lipoproteic barrier built by the antiluminal and the basement membranes. The results we obtained are summarized in Table 1.

TABLE 1 Activities of several drug-metabolizing enzymatic systems in confluent primary cultures (15-day-old) of rat brain cerebrovascular cells.

Substrate	Enzymatic system	Activity (nmol/min·mg protein)
1.5 μM Ethoxyresorufin	CYP1A	Not detectable
1.5 μM Benzoxyresorufin	CYP1A + CYP2B	Not detectable
10 μM Cytochrome *c*	NADPH-P450 reductase	8.98 ± 1.85 (n=6)
0.5 mM 1-Naphthol	UGT 1*6	11.70 ± 2.70 ‡ (n = 6)
0.25 mM 1-Naphthol	Phenol sulfotransferase	Not detectable

‡: native (without detergent activation) activity.

It seems that enzymatic activities which are expressed during the early steps of cerebral development, like NADPH-cytochrome P450 reductase (Ghersi-Egea et al., 1989) and 1-naphthol-UGT (Gradinaru, 1995; Leininger-Muller et al., in preparation) are also rapidly expressed in primary cultures of cerebrovascular endothelial cells and are therefore easily measurable. Conversely, activities which are practically undetectable in the brains of fetal or newborn rats, like CYP1A or phenol sulfotransferase, present also very low levels in these cultured cells, even if they are efficiently active in microvessels isolated from adult rat brains. Therefore, cerebrovascular endothelial cells in primary cultures are useful tools for the study of NADPH-cytochrome P450 reductase-dependent free radicals-induced damage to the blood-brain barrier (Lagrange et al., 1996), as well as for the UGT isoforms identification, control of expression and activity measurements (Gradinaru, 1995).

PROBLEM SOURCES AND 'QUALITY CONTROL'

One of the main problems of preparations of primary cultures remains, of course, the risk of contamination by undesirable microorganisms. After decapitation, the head of each rat was rapidly immersed in ethanol and the brain was removed using sterilized tools. All culture media should be carefully sterilized by ultrafiltration and the culture cells are submitted to sterilization by UV illumination before seeding.

The second problem which may appear during primary cultures of endothelial cells is the development of other cells, like pericytes or fibroblasts. The selective destruction of these cells, or the removing and transplant of endothelial cells on other culture wells, could be a good solution for obtaining pure cultures.

CONCLUDING REMARKS

There are presently numerous possible utilisations of cultured cells as models of the functioning in vivo, allowing also for instance toxicity or mutagenesis studies. The evaluation of chemically-induced toxicity, especially in a heterogenous system such as the brain, is rather difficult in the intact animal because numerous factors (e.g. hormonal status, stress, metabolism, plasmatic binding) are not entirely under experimental control. Hence a simplified model, such as cerebrovascular endothelial cells primary cultures, affords many advantages over in vivo techniques as transport, metabolism and formation of reactive metabolites can be easily studied. We describe here the first study concerning the enzymatic capacity of rat cerebrovascular endothelial cells towards xenobiotics. The main result of drug metabolizing enzyme activities is an increase of the polarity, therefore the water solubility, of drugs. This allows in the body an easier clearance of foreign molecules by urine or bile, nevertheless, the formation of watersoluble metabolites in the central nervous system is rather a difficulty, as these metabolites may encounter difficulties in leaving the brain through the BBB. For instance, cocaine and its metabolites remain in the brain significantly longer than in the blood (Madden and Powers, 1990) and a single intake of phencyclidine may produce behavioral disorders lasting several weeks, as a result of the high lipophilicity of the molecule and of some of its neuroactive metabolites (Misra *et al.*, 1979). However, the presence in endothelial cells of specific, probably active, transport systems allowing the efflux of the hydrophilic drug metabolites has been demonstrated for glucuronides (Leininger *et al.*, 1991), and postulated for glutathione conjugates (Minn *et al.*, 1993).

The protective function of these enzymatic activities could explain the relatively high resistance of rats to the neurotoxicity of MPTP, as rat brain capillaries contain high MAO-B activities, able to metabolize MPTP to MPP^+ at a rate sufficient to prevents its entry into the brain (Riachi *et al.*, 1990). However, as a more general rule, the activities of the enzymatic systems localized in the BBB are rapidly saturable, thus sufficiently large concentrations of a molecule will 'force' it past the BBB. This could result in toxic effects but can also be used for therapeutic purposes, for instance with L-DOPA in Parkinson's disease therapy.

Cerebrovascular endothelial cells alteration may endanger normal brain functioning by a loss of their protective properties. We already know that oxygen related free radicals, generated during the metabolism of some molecules of foreign origin, may increase the monocellular layer permeability to sucrose (Lagrange *et al.*, 1996). The production of oxygen-derived free radicals represents a potential hazard for the capacity of specific transport systems for neurotransmitter precursors and brain nutrients (Pardridge, 1983) and for the regulation of cerebral blood flow, as lipid peroxidation decreases endothelial membrane fluidity. Fortunately, several enzymatic mechanisms have been demonstrated to be present in brain microvessels, protecting them against the continuous enzymatic production of oxygenated reactive metabolites (Tayarani *et al.*, 1987).

ACKNOWLEDGEMENTS

The financial support from the Commission of the European Communities (BMH1-CT92-1193) is gratefully acknowledged. D. Gradinaru is a research fellow from the Sandoz Foundation for Gerontological Research.

REFERENCES

Baranczyk-Kuzma, A., Audus, K.L. and Borchardt, R.T. (1989) Substrate specificity of phenol sulfotransferase from primary cultures of bovine brain microvessel endothelium. *Neurochem. Res.*, **14**, 689–690.
Bates, I.P. (1985) Permeability of the blood-brain barrier. *Trends Pharmacol. Sci.*, **6**, 447–450.
Bayol-Denizot, C., Lagrange, P., Chat, M. and Minn, A. (1996) Superoxide radical formation by primary cultures of rat brain astrocytes, neurons and cerebrovascular endothelial cells (submitted).
Campbell, M.E., Grant, D.M., Inaba, T. and Kalow, W. (1987) Biotransformation of caffeine, paraxanthine, theophylline and theobromine by polycyclic aromatic hydrocarbon-inducible cytochromes P-450 in human liver microsomes. *Drug Metab. Dispos.*, **15**, 237–249.

Chen, Z.R., Irvine, R.J., Bochner, F. and Somogyi, A.A. (1990) Morphine formation from codeine in rat brain: A possible mechanism of codeine analgesia. *Life Sci.*, **46**, 1067–1074.

De Vries, H.E., Kuiper, J., De Boer, A.G., van Berkel, Th.J.C. and Breimer, D.D. (1993) Characterization of the scavenger receptor on bovine cerebral endothelial cells *in vitro. J. Neurochem.*, **61**, 1813–1821.

Dutton, G.J. (1980) *Glucuronidation of drugs and other compounds.* CRC Press, Boca Raton, Florida, pp. 65–158.

Ghersi-Egea, J.F., Leininger-Muller, B., Suleman, G., Siest, G. and Minn, A. (1994) Localization of several drug-metabolizing enzymes activities in blood-brain interface structures. *J. Neurochem.*, **62**, 1089–1096.

Ghersi-Egea, J.F., Minn, A., Daval, J.L., Jayyozi, Z., Arnould, V., Souhaili-el Amri, H. and Siest, G. (1989) NADPH: Cytochrome P-450(c) reductase: biochemical characterization in rat brain and cultured neurones and evolution of activity during development. *Neurochem. Res.*, **14**, 883–888.

Ghersi-Egea, J.F., Minn, A. and Siest, G. (1988) A new aspect of the protective functions of the blood brain barrier: Activities of four drug-metabolizing enzymes in isolated brain microvessels. *Life Sci.*, **42**, 2515–2523.

Ghersi-Egea, J.F., Tayaranni, Y., Lefauconnier, J.M. and Minn, A. (1988) Enzymatic protection of the brain: role of 1-naphthol UDP-glucuronosyltransferase from cerebral tissue and cerebral microvessels. In G. Siest, J. Magdalou and B.Burchell (Eds.), *Cellular and Molecular Aspects of Glucuronidation.* Colloque INSERM/John Libbey Eurotext Ltd, London, **173**, 169–175.

Gradinaru, D. (1995) Ageing and reactive oxygen species-induced modifications in rat brain glucuronidation process. Research report to the Sandoz Foundation for Gerontological Research, p. 26.

Haglund, L., Kohler, C., Haaparenta, Y., Goldstein, M. and Gustafsson, J.A. (1984) Presence of NADPH-cytochrome P-450 reductase in central catecholaminergic neurones. *Nature*, **307**, 259–262.

Hannsson, T., Tindberg, N., Inglman-Sundberg, M., Köhler C. (1990) Regional distribution of ethanol-inducible cytochrome P-450 IIE1 in the rat central nervous system. *Neuroscience*, **34**, 451–453.

Hu, Z.Y., Bourreau, E., Jung-Testas, I., Robel, P. and Baulieu, E.E. (1987) Neurosteroids: Oligodendrocytes mitochondria convert cholesterol to pregnenolone. *Proc. Natl. Acad. Sci. USA*, **85**, 8215–8219.

Joó, F. (1985) The blood-brain barrier *in vitro*: Ten years of research on microvessels isolated from the brain. *Neurochem. Int.*, **7**, 1–25.

Kanda, T., Yamawaki, M., Ariga, T. and Yu, R.K. (1995) Interleukin 1β up-regulates the expression of sulfoglucuronosyl paragloboside, a ligand for L-selectin, in brain microvascular endothelial cells. *Proc. Natl. Acad. Sci. USA*, **92**, 7897–7907.

Karoum, F., Chuang, L.-W. and Wyatt, R.J. (1993) Biochemical and pharmacological characteristics of conjugated catecholamines in the rat brain. *J. Neurochem.*, **40**, 1735–1741.

Lagrange, P., Livertoux, M.H., Ghersi-Egea, J.F., Grassiot, M.C. and Minn, A. (1994) Superoxide anion formation during monoelectronic reduction of xenobiotics by preparations of rat brain cortex, microvessels and choroid plexus. *Free Rad. Biol. Med.*, **17**, 355–359.

Lagrange, P., Romero, I.A., Minn, A. and Revest, P.A. (1996) Transendothelial permeability changes induced by xenobiotic-derived free radicals in an *in vitro* model of the blood-brain barrier. (submitted).

Le Gascogne, C., Gouézou, M., Robel, P., Defaye, G., Chambaz, E., Waterman, M. and Baulieu, E.E. (1989) The cholesterol side-chain cleavage complex in human brain white matter. *J. Neuroendocrin.*, **1**, 153–156.

Leininger, B., Ghersi-Egea, J.F., Siest, G. and Minn, A. (1991) *In vivo* study of the elimination from brain tissue of an intracerebrally-formed xenobiotic metabolite, 1-naphthyl-β-D-glucuronide. *J. Neurochem.*, **56**, 1163–1168.

Leininger-Muller, B., Gradinaru, D., Suleman, G.F., Daval, J.L. and Minn, A. Cellular distribution and developmental evolution of 1-naphthol glucuronidation in rat brain. (submitted).

Lemoine, A., Gautier, J.C., Azoulay, D., Kieffel, L., Belloc, C., Guengerich, F.P., Maurel, P., Beaune, P. and Leroux, J.P. (1993) Major pathway of imipramine metabolism is catalyzed by cytochromes P450 1A2 and P450 3A4 in human liver. *Mol. Pharmacol.*, **43**, 827–832.

Madden, J.A. and Powers, R.H. (1990) Effect of cocaine and cocaine metabolites on cerebral arteries *in vitro. Life Sci.*, **47**, 1109–1114.

Majewska, M.D. and Schwartz, R.D. (1987) Pregnenolone-sulfate: An endogenous antagonist of the gamma-aminobutyric receptor complex in brain? *Brain Res.*, **404**, 355–360.

Meek, J.L. and Neef, N.H. (1972) Fluorometric estimation of 4-hydroxy-3-methoxyphenyl-ethyleneglycol sulfate in brain. *Brit. J. Pharmacol.*, **45**, 435–444.

Minn A., Ghersi-Egea, J.F., Perrin, R., Leininger-Muller, B. and Siest, G. (1991) Drug metabolizing enzymes in the rat brain and brain microvessels. *Brain Res. Rev.*, **16**, 65–82.

Minn, A., Leininger-Muller, B., Ghersi-Egea, J.F., Suleman, G., Grassiot, M.C. and. Siest, G. (1993) Penetration, biotransformation et elimination des metabolites de medicaments dans la barriere hematoencephalique. *Circul. Metab. Cerveau (Paris)*, **10**, 117–131.

Misra, A.L., Pontani, P.B. and Bortolomeo, J. (1979) Persistence of phencyclidine (PCP) and metabolites in rat brain and adipose tissue, and implication for long-lasting behavioral effects. *Res. Comm. Chem. Pathol. Phar.*, **3**, 431–445.

Oldendorf, W.H. (1977) The blood-brain barrier. *Exp. Eye Res.*, **25**, (Suppl.) 177–190.

Pardridge, W.M. (1983) Transport of nutrients and hormones through the blood-brain barrier. *Fed. Proc.*, **43**, 201–204.

Pasanen, M., Pellinen, P., Stenbàck, F., Juvonen, R.O., Raunio, H. and Pelkonen, O. (1995) The role of CYP enzymes in cocaine-induced liver damage. *Arch. Toxicol.*, **69**, 287–290.

Perrin, R., Minn, A., Ghersi-Egea, J.F., Grassiot, M.C. and Siest, G. (1990) Distribution of cytochrome P-450 activities towards alkoxyresorufin derivatives in rat brain regions, subcellular fractions and isolated cerebral microvessels. *Biochem. Pharmacol.*, **40**, 2145–2151.

Riachi, N.J., LaManna, J.C. and Harik, S.I. (1989) Entry of 1-methyl-4-phenyl-1,2,5,6-tetrahydropyridine into the rat brain. *J. Pharmacol. Exp. Ther.*, **249**, 744–748.

Smith, W.L. (1989) The eicosanoids and their biochemical mechanism of action. *Biochem. J.*, **259**, 315–324.

Suleman, F.G., Leininger-Muller, B., Ghersi-Egea, J.F. and Minn, A. (1993) UDP-glucuronosyl-transferase activities in rat brain microsomes. *Neurosci. Lett.*, **161**, 219–222.

Tayarani, I., Chaudiere, J., Lefauconnier, J.M. and Bourre, J.M. (1987) Enzymatic protection against peroxidative damage in isolated brain capillaries. *J. Neurochem.*, **48**, 1399–1402.

Walther, B., Ghersi-Egea, J.F., Minn, A. and Siest, G. (1987) Brain mitochondrial cytochrome P-450scc: Spectral and catalytic properties. *Arch. Biochem. Biophys.*, **254**, 592–596.

II.5. IMMUNOHISTOCHEMICAL AND ELECTRONMICROSCOPY DETECTIONS

M.A. DELI, C.A. SZABÓ, N.T.K. DUNG AND F. JOÓ[†]

Laboratory of Molecular Neurobiology, Institute of Biophysics,
Biological Research Center of the Hungarian Academy of Sciences
P.O. Box 521, H-6701 Szeged, Hungary

Primary cultures of endothelial cells were obtained from cortical grey matter of two-week-old rats using a two-step enzymic digestion followed by a Percoll gradient centrifugation. Rat cerebral endothelial cells (RCECs) formed a monolayer of spindle-shaped, tightly attached cells on rat tail collagen matrix. RCECs specifically stained for Factor VIII (FVIII) and alkaline phosphatase (AP), and bound *Bandeiraea simplicifolia* isolectin I-B$_4$ (BS-I-B$_4$). At electron microscopic level gap junction-like attachment sites were observed. Contaminating cells in the cultures were eliminated by complement mediated cytolysis using anti-Thy 1.1 antibody. On average, a value of 120 $\Omega.cm^2$ for transendothelial electrical resistance (TEER) was measured for RCEC monolayers. Primary cultures of RCECs were used to study regulatory enzymes in signal transduction.

INTRODUCTION

With the recognition that most of the endothelial cells resisted damage during isolation, remained viable, and could be maintained in tissue culture conditions (Panula *et al.*, 1978), a new generation of blood-brain barrier (BBB) model systems was developed, which seemed to be devoid of most of the problems experienced with the isolated cerebral microvessels. Several procedures and many modifications have been worked out for culturing RCECs in different laboratories since then (Phillips *et al.*, 1979; Spatz *et al.*, 1980; Bowman *et al.*, 1981; Diglio *et al.*, 1982; Rupnick *et al.*, 1988; Gordon *et al.*, 1991; Dux *et al.*, 1991; Abbott *et al.*, 1992). Except that of Phillips *et al.* (1979) all methods used grey matter as a starting material, collagenase and/or collagenase-dispase for enzymic digestion. In most of the protocols endothelial cells were separated by a Percoll gradient centrifugation step. In our method, which is very close to the method established in the laboratory of N.J. Abbott (see chapter II.1), first the microvessels were separated from brain tissue with a collagenase digestion, and then basement membrane and the majority of perivascular cells were removed by a second enzymic treatment. Endothelial cell clusters were further purified by Percoll gradient centrifugation before plating. Optimal purity of RCEC cultures was obtained by selectively lysing Thy 1.1 expressing contaminating cells.

[†]Deceased.

MATERIALS AND METHODS

Animals

Two-week-old CFY rats of either sex were obtained from the Institute's animal house.

Chemicals

Products	Supplier and catalogue No.
Establishment of primary cultures	
Collagenase type CLS2	Worthington, NJ, USA
Collagenase-dispase	Boehringer 1097113
Percoll	Pharmacia 17-0891-01
Bovine serum albumin (BSA)	Sigma A 2153
L-glutamine	Gibco 15039-019
Dulbecco's modified Eagle's medium (DME)	Sigma D 5648
Dulbecco's modified Eagle's medium nutrient mixture F-12 Ham (DME/F-12)	Sigma D 8900
Penicillin-Streptomycin	Gibco 15145-014
Gentamicin	Gibco 15710-031
Fetal calf serum (FCS)	HyClone and Sigma
Complement mediated specific cytolysis	
monoclonal anti-mouse Thy 1.1 antibody	Sigma M 7898
complement serum HLA-ABC, rabbit	Sigma S 7764
Characterisation	
5-bromo-4-chloro-3-indolyl phosphate (BCIP)	Sigma B 6149
nitro blue tetrazolium (NBT)	Sigma N 6876
Bandeiraea simplicifolia isolectin B$_4$	Sigma L 5391
3,3'-diaminobenzidine (DAB)	Sigma D 5637
Fluorescein isothiocyanate-dextran (FITC-dextran)	Sigma FD-70S

Establishment of Primary Cultures

Dissection of brains

Rats were anaesthetised with ether. After a thorough rinse with 70% ethanol and then with iodine in 70% ethanol, the heads were cut, and placed in a sterile glass petri dish. In the laminar flow box forebrains were removed from the skulls (without the cerebellum) using sterile microdissecting forceps and scissors, and collected in cold sterile phosphate buffered saline (PBS, without calcium and magnesium, pH 7.4). Meninges were removed on sterile filter paper (Whatman 3M) from each brain hemisphere and at the same time white matter was 'peeled off' with the aid of fine curved forceps. Grey matter was carefully collected from the filter paper (meninges tend to stick to it) and minced to approximately 1 mm^3 pieces using sterile disposable scalpels in the first incubation medium (3 mg/ml collagenase CLS2, 1 mg/ml BSA in DME containing antibiotics) in a sterile glass petri dish.

Enzyme digestions

The minced tissue was transferred into a centrifuge tube (35 ml, Oakridge-type with screw cap) with the rest of the collagenase solution (total: 15 ml/10 brains) and triturated with a pipette (10 up and down), and then incubated at 37 °C for 1.5 h in a shaking waterbath. After this incubation, 25 ml of cold 25% BSA-DME was added to the homogenate, mixed well by trituration and centrifuged at 1000 g for 20 min. The myelin layer and the supernatant was aspirated, the pellet washed once in DME (1000 g for 10 min) and then further incubated in the waterbath for a maximum of 2 h in 10 ml of the second incubation medium containing 1 mg/ml collagenase-dispase in DME.

Percoll gradient centrifugation

The cell suspension was centrifuged (700 g for 5 min). The pellet was suspended in 2 ml DME and carefully layered onto a continuous 33% Percoll gradient and centrifuged at 1000 g for 10 min. For the gradient 10 ml Percoll, 18 ml PBS, 1 ml FCS and 1 ml 10 × concentrated PBS were mixed, sterile filtered and centrifuged at 4 °C, 30000 g for 1 h.

Plating and feeding cells

The band of the endothelial cell clusters (clearly visible as a white-greyish layer above the red blood cells) was aspirated and washed twice in DME (1000 g, 10 min). The cells were then suspended in culture medium (DME/F-12 containing 100 U/ml penicillin, 100 μg/ml streptomycin, 50 μg/ml gentamicin, 2 mM glutamine, 20% heat inactivated FCS) and were seeded onto rat tail collagen-coated 35 mm plastic dishes or 25 mm cell culture inserts (seeding density 10 cm^2/brain). The medium was changed the next day, later on, every third day.

Yield

Starting with the culture from 10 brains, we could obtain confluent primary culture of RCECs in 10 pieces of 35 mm tissue culture dish, equivalent approximately to 100 cm^2 surface area.

Comments

Incubation media for enzymic digestions were always prepared freshly with lyophilised enzymes, and then sterilised by filtration. Their pH was adjusted to 7.4. During the separation we used only plasticware. If it was necessary to use any kind of glassware, we coated it before use with BSA.

SPECIFIC APPLICATION

For RCEC monolayers, passage 1–2, grown on Falcon 25 mm inserts, 120 Ω.cm^2 TEER was obtained. The passage of 70 kDa FITC-dextran was restricted through RCECs: 99.12 ± 8.79 $\mu g/cm^2/h$ *vs.* 655.80 ± 12.37 $\mu g/cm^2/h$ ($n = 6$) in the case of cell-free filters.

Using cells grown on collagen-coated dishes and/or coverslips, the production, presence, and phosphorylation of the α-subunit of calcium/calmodulin-stimulated protein kinase II have been described in primary cultures of RCECs (Deli *et al.*, 1993). The expression of seven different protein kinase C isoforms using reverse transcriptase-polymerase chain reaction was investigated in cerebral endothelial cells (Krizbai *et al.*, 1995). The expression patterns of brain tissue, isolated microvessels, RCECs, an immortalized cerebral endothelial cell line and aortic endothelial cells were determined and analysed.

PROBLEM SOURCES AND QUALITY CONTROL

Characterisation of the Cultures

Factor VIII-related antigen (FVIII) immunohistochemistry

After a brief washing in PBS and fixing in ethanol at 4 °C for 15 min, cells were treated with 1% H_2O_2 in PBS for 10 min, followed by washing in PBS. Non-specific binding sites were blocked by incubation in 3% normal goat serum in PBS at room temperature for 20 min. Anti-FVIII rabbit immunoglobulin (Dako) was used as primary antibody and biotin-labelled anti-rabbit IgG (Dako) as secondary antibody, both applied for 30 min. After a 30 min incubation with avidin-biotin-horseradish peroxidase (HRP) complex (ABC kit, Vector Lab, CA, USA), DAB was used as HRP substrate, followed by hematoxylin-eosin (HE) counterstaining.

Lectin binding

RCECs were washed in PBS, fixed in 4% formalin and 70% ethanol in PBS for 15 min, treated with 1% H_2O_2 in PBS for 10 min, washed again in PBS and then incubated in 15 $\mu g/ml$ HRP-conjugated BS-I-B$_4$ in 0.1% BSA-PBS for 90 min. DAB was used as HRP substrate. The preparations were counterstained with HE.

AP histochemistry

RCECs were washed in PBS, fixed in 2% paraformaldehyde-PBS for 2 min, washed again in PBS, and then incubated in 0.41 mM NBT, 0.40 mM BCIP in Tris buffered saline, pH 9.5 for 3 h. The reaction was stopped by washing in PBS, and subsequently subjected to graded dehydration and mounted in Entellan®.

Transmission electron microscopy

RCEC cultures were fixed for 30 min in long Karnovsky's fixative and then washed in 0.1 M phosphate buffer, pH 7.4. Cells were scraped off the surface, and centrifuged in the buffer. The pellet was post-fixed in 1% OsO_4, dehydrated and embedded in Spurr resin. Thin sections were stained in lead citrate and finally examined in a Zeiss 902 electron microscope.

Endothelial Cells

The small vessel fragments obtained at the end of the isolation procedure attached rapidly to collagen-coated surfaces, and in 2–3 days colonies of RCECs emerged (Figure 1a), and formed a non-overlapping continuous monolayer at the end of the first week (Figure 1c) with some swirling patterns. RCECs displayed a so called 'fibroblast-like' morphology: cell-shape was fusiform with an oval nuclei in the centre with neighbouring cells tightly apposed to each other. Cells gave specific immunohistochemical staining with anti-FVIII antibody (Figures 1b and d), bound the galactose-specific BS-I-B_4 isolectin (Figure 1f) and showed positive histochemical staining for AP. Cells were abundant in mitochondria and endoplasmic reticulum (Figures 2a and b). Some attachment sites were without specialisation (Figure 2c), while others were gap junction-like (Figure 2d).

Contaminating Non-Endothelial Cells

In primary cultures of RCECs the presence of contaminating non-endothelial cells could be occasionally observed (Figures 3a and c), which did not express FVIII (Figure 1d). These cells, mostly pericytes and very few astrocytes, expressed Thy 1.1 antigen and therefore could be removed by selective cytolysis using anti-Thy 1.1 antibody and complement (Risau *et al.*, 1990). For detailed description of the method see chapters A.II.1. and II.2. We have followed the method established in the laboratory of N.J. Abbott, and could considerably reduce the number of contaminating cells (Figure 3).

CONCLUDING REMARKS

The *in vitro* BBB model systems are applicable not only to biochemistry and physiology but also to drug research, and may contribute to the improvement of the transport of substances through the BBB (Joó, 1992). The *in vitro* approach has

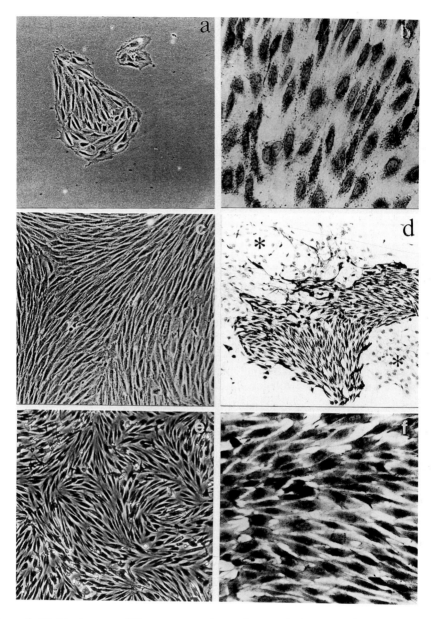

Figure 1 Light microscopy. Phase contrast photomicrographs from cerebral endothelial cells (a, c and e). A colony of primary RCECs 3 days after seeding (a), and a confluent culture 7 days after seeding (c). Confluent primary culture of PCECs, 7 days after seeding (e), magnification 200 ×. FVIII immunohistochemistry, HE counterstaining (b and d). Specific, perinuclear cytoplasmic dot-like staining of RCEC (b), 400 ×. An islet of RCECs was well stained for FVIII (d), while surrounding contaminating cells were not (asterisks), 80 ×. *Bandeiraea simplicifolia* isolectin B_4 binding to RCEC (f), 400 ×.

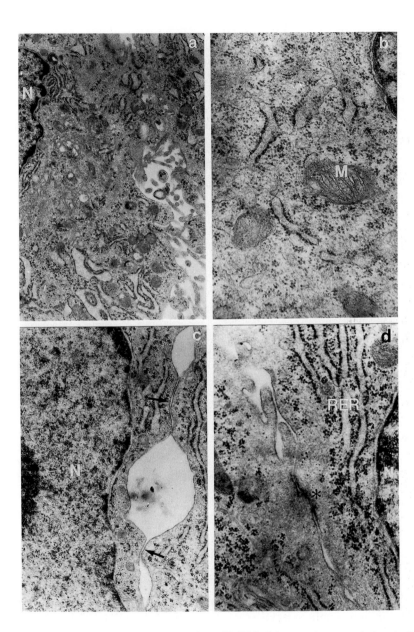

Figure 2 Electron microscopy. Low power view of RCECs (a), note the extended endoplasmic reticulum, numerous mitochondria; N = nucleus, magnification 15000 ×. High magnification of the cytoplasm (b); M = mitochondrion, 32000 ×. Attachment sites (arrows) of two RCECs without specialization (c); N = nucleus, 26000 ×. Gap junction-like specialized membrane attachment (asterisk) could be seen between two RCECs (d); RER = rough endoplasmic reticulum, N = nucleus, 38000 ×.

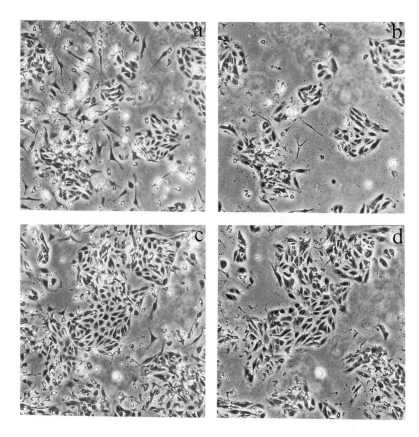

Figure 3 Selective cytolysis of contaminating cells in primary cultures of RCEC by anti-Thy 1.1 antibody and complement. Phase contrast microscopy pictures from primary cultures of RCECs 3 days after seeding, magnification 200 ×. Cultures before (a and c), and immediately after the treatment (b and d).

been and should remain an excellent model of the BBB to help unravel the complex molecular interactions underlying and regulating the permeability of the cerebral endothelium.

It should be stressed, however, that studies on permeability and transport across RCEC monolayers ought to be carried out in cultures having parameters resembling those *in vivo*, e.g. monolayers with TEER equal or higher than 500 $\Omega.cm^2$. Methods that co-cultivate cerebral endothelial cells with astrocytes (Dehouck *et al.*, 1990) and/or treatment of monolayers with compounds that increase intracellular cyclic AMP concentration, thereby enhance the 'tightness' of the tight junctions between endothelial cells (Rubin *et al.*, 1991; Deli *et al.*, 1995) could provide means to reach that goal.

REFERENCES

Abbott, N.J., Hughes, C.C.W., Revest, P.A. and Greenwood J. (1992) Development and characterisation of a rat brain capillary endothelial culture: Towards an *in vitro* blood-brain barrier. *J. Cell Sci.*, **103**, 23–37.

Bowman, P.D., Betz, A.L., Ar, D., Wolinsky, J.S., Penney, J.B., Shivers, R.R. and Goldstein, G.W. (1981) Primary culture of capillary endothelium from rat brain. *In Vitro*, **17**, 353–362.

Dehouck, M.-P., Méresse, S., Delorme, P., Fruchart, J.C. and Cecchelli, R. (1990) An easier, reproducible and mass-production method to study the blood-brain barrier *in vitro*. *J. Neurochem.*, **57**, 1798–1801.

Deli, M.A., Joó, F., Krizbai, I., Lengyel, I., Nunzi, M.G. and Wolff, J.R. (1993) Calcium/calmodulin-stimulated protein kinase II is present in primary cultures of cerebral endothelial cells. *J. Neurochem.*, **60**, 1960–1963.

Deli, M.A., Dehouck, M-P., Ábrahám, C.S., Cecchelli, R. and Joó, F. (1995) Penetration of small molecular weight substances through cultured bovine brain capillary endothelial cell monolayers: the early effects of cyclic adenosine 3′,5′-monophosphate. *Exp. Physiol.*, **80**, 675–678.

Diglio, C.A., Grammas, P., Giacomelli, F. and Wiener, J. (1982) Primary culture of rat cerebral microvascular endothelial cells: Isolation, growth and characterization. *Lab. Invest.*, **46**, 554–563.

Dux, E., Noble, L. and Chan, P.H. (1991) Glutamine stimulates growth in rat cerebral endothelial cell culture. *J. Neurosci Res.*, **29**, 355–361.

Gordon, E.L., Danielsson, P.E., Nguyen, T.S. and Winn, H.R. (1991) A comparison of primary cultures of rat cerebral microvascular endothelial cells to rat aortic endothelial cells. *In Vitro*, **27A**, 312–326.

Joó, F. (1992) Cerebral microvessels in culture, an update. *J. Neurochem.*, **58**, 1–17.

Panula, P., Joó, F. and Rechardt, L. (1978) Evidence for the presence of viable endothelial cells in cultures derived from dissociated rat brain. *Experientia*, **34**, 95–97.

Phillips, P., Kumar, P., Kumar, S. and Woghe, M. (1979) Isolation and characterization of endothelial cells from rat and cow brain white matter. *J. Anat.*, **129**, 261–272.

Risau, W., Engelhardt, B. and Wekerle, H. (1990) Immune function of the blood-brain barrier: Incomplete presentation of protein (auto-)antigens by rat brain microvascular endothelium *in vitro*. *J. Cell Biol.*, **110**, 1757–1766.

Rubin, L.L., Hall, D.E., Parter, S., Barbu, C., Cannon, C., Horner, H.C., Janatpour, M., Liaw, C.W., Manning, K., Morales, J., Tanner, L.I., Tomaselli, K.J. and Bard, F. (1991) A cell culture model of the blood-brain barrier. *J. Cell Biol.*, **115**, 1725–1735.

Rupnick, M.A., Carey, A. and Williams, S.K. (1988) Phenotypic diversity in cultured cerebral microvascular endothelial cells. *In Vitro*, **24**, 435–444.

Spatz, M., Bembry, J., Dodson, R.F., Hervonen, H. and Murray, M.R. (1980) Endothelial cell cultures derived from isolated cerebral microvessels. *Brain Res.*, **191**, 577–582.

III.1. ISOLATION AND PRIMARY CULTURES OF BOVINE BRAIN MICROVESSEL ENDOTHELIAL CELLS (BMEC)

A.G. de BOER, P.J. GAILLARD, H.E. de VRIES AND D.D. BREIMER

Division of Pharmacology, Leiden/Amsterdam Center for Drug Research (LACDR), University of Leiden, P.O. Box 9503, 2300 RA Leiden, The Netherlands

The use of brain microvessels and the isolation and culture of bovine brain microvessel endothelial cells are described. Pitfalls in the application and culture of these cells are also explained.

INTRODUCTION

In vivo and *in vitro* techniques have contributed a great deal towards research of drug transport across the blood-brain barrier (BBB). The *in vivo* techniques allow the estimation of the rates and quantification of drugs transported across the BBB, while a more detailed qualitative and quantitative studies of BBB transcytosis processes can be carried out *in vitro*. Unfortunately the results generated by the two techniques may differ considerably. This is partly due to some intrinsic properties of some *in vitro* models. The possibilities and limitations for the use of some of the *in vitro* BBB models are discussed in chapter II.2 "Possibilities, Limitations and Isolation Procedures of Rat Brain Microvessel-Endothelial Cell Culture Techniques". The materials and procedures for the isolation of bovine brain microvessel endothelial cells (BMEC) are described below.

ISOLATION AND PRIMARY CULTURES OF CALF AND BOVINE BRAIN MICROVESSEL ENDOTHELIAL CELLS (BMEC) (Rubin *et al.*, 1991, see Part A, II.2)

Reagents and Materials

Reagents

Dulbecco's Modified Eagle's Medium	Sigma D-1152
MEM non-essential amino acid solution (100 ×)	Sigma M-7145
Endothelial cell trypsin	Sigma T-4299
Poly-D-lysine	Sigma P-7280
DMSO	Baker Analyses Reagents
Heparin	Sigma H-3125
Streptomycin sulphate	Sigma S-9137
Penicillin G sodium	Sigma P-3032
L-glutamine	Biochemical

Bovine serum albumin	Sigma
Collagenase CLS3	Worthington Biochem. Corp.
Trypsin TRL	Worthington Biochem. Corp.
Deoxyribonuclease I (DNase)	Worthington Biochem. Corp.
Foetal bovine serum	Gibco 10090–075
Collagen Type VII (from rat tail, acid soluble)	Sigma C-8897
Inorganic salts	Baker Analyses Reagents
HEPES	Sigma H-9136
CPT-cAMP	Sigma C-3912
RO-20-1724	Calbiochem 557502
Trypsin-EDTA (1X)	Gibco 043-05300 M
Transferrin	Sigma T-0519
Putrescine	Sigma P-7505
Sodium selenite	Sigma S-5261
Versene	Gibco 15040-033
Fibronectin	Boehringer Mannheim 1080 938

Materials for isolation of endothelial cells from one brain

Numbers on the left refer to the quantity required:

 1 40 ml Wheaton homogeniser + pestle A and pestle B (autoclaved).
 2 sterile cloths for the clean bench.
 1–2 Millipore (Svinnex 47) filter holder (autoclaved).
 2 225 ml Falcon flasks (sterile).
 3 50 ml Falcon flasks (sterile).
 4 15 cm petri dishes.
 2–4 9 cm petri dishes.
 1 scalpel (sterile).
 1–2 razor blades (autoclaved).
 15 150 μm filters (autoclaved).
 3 sterile gloves.
 1 20 ml syringe (sterile).
 5 coated growth-flasks (75 cm^2).
 2 boxes with ice.

Preparation of Solutions and Media

1. DMEM (Dulbecco's modified Eagle's medium), pH 7.4

3.7 g	NaHCO$_3$		
0.10 g	Streptomycin sulphate	100	μg/ml
0.10 g	Penicillin G sodium	100	μg/ml
1000 ml	Milli-Q water (Millipore)		
1 bottle	DMEM		

– Weigh streptomycin sulphate and penicillin G
– Dissolve in 1 l Milli-Q water and adjust pH with 1.0N NaOH from ± pH 5.1 (DMEM) to pH 7.2.
– Sterile filtration in a laminar flow cabinet. The pH will then increase to 7.4.

2. PBS buffer (phosphate buffered saline)

0.455 g	$NaH_2PO_4 \cdot H_2O$	3.3 mM
1.191 g	$Na_2HPO_4 \cdot 2H_2O$	6.7 mM
8.5 g	NaCl	150 mM
1000 ml	Milli-Q water (Millipore)	

– Weigh the inorganic salts.
– Dissolve in Milli-Q water.
– Adjust pH with 1.0 N NaOH to pH 7.4.
– Autoclave.
– Store in the refrigerator.

3. Filter growth medium (N2)

75 μl	Transferrin solution (20 mg/ml)	10 μg/ml
75 μl	Putrescine solution (200 mM)	100 μM
15 μl	Sodium selenite solution (0.3 mM)	30 nM
150 ml	DMEM + 10% FCS	

– Mix the solutions.
– Sterile filtration in a laminar flow cabinet.
– Store in the refrigerator.

4. N2/ACM (1:1) + cAMP/RO-20-1724

100 μl	CPT-cAMP (25 mM)	250 μM
50 μl	RO-20-1724 (35 mM)	17.5 μM
50 ml	N2	
50 ml	Astrocyte conditioned medium (ACM)	

– Mix the solutions under aseptic conditions.
– Store in the refrigerator.

5. Growth medium

50%	DMEM + 10% FCS
50%	ACM
25 μl/ml	0.5% heparin

– Mix all solutions.
– Store in the refrigerator.

6. DMEM + 10% foetal calf serum (FCS)

50 ml   FCS
500 ml   DMEM

 – Deactivate FCS for 30 min at 56 °C.
 – Add FCS to MEM under sterile conditions.
 – Store in the refrigerator.

7. Freeze mix

1 ml   DMSO   10%
9 ml   FCS   90%

8. Collagenase solution

X U/mg   Collagenase CLS3   2098 U/ml
10 ml   PBS buffer

 – Weigh the calculated amount of collagenase needed.
 – Dissolve in PBS buffer.
 – Sterile filter in a laminar flow cabinet.
 – Store in 2 ml aliquots at −20 °C.

9. Trypsin solution

X U/mg   Trypsin TRL   907 U/ml
10 ml   PBS buffer

 – Weigh the calculated amount of trypsin needed.
 – Dissolve in PBS buffer.
 – Store in 2 ml aliquots at −20 °C.

10. DNase solution

X U/mg   DNase I   3398 U/ml
5 ml   PBS buffer

 – Weigh the calculated amount of DNase I needed.
 – Dissolve in PBS buffer.
 – Store in 1 ml aliquots at −20 °C.

11. Digest mix

2 ml   Collagenase solution   210 U/ml
2 ml   Trypsin solution   91 U/ml
1 ml   DNase I solution   170 U/ml
15 ml   DMEM + 10% FCS

- Thaw collagenase-, trypsin- and DNase-aliquots.
- Mix the solutions under sterile conditions.
- Store at −20 °C or use instantly.

12. 0.5% Heparin (5 mg/ml)

125 mg   Heparin
25 ml   Milli-Q water (Millipore)

- Weigh heparin.
- Dissolve in 25 ml milli-Q water.
- Store in the refrigerator.

13. Transferrin (20 mg/ml)

20 mg   Transferrin   20 mg/ml
1 ml   Milli-Q water (Millipore)

- Weigh transferrin and dissolve in Milli-Q water.
- Store in 100 μl aliquots at −20 °C.

14. Putrescine (200 mM)

32 mg   Putrescine (MW = 161.1)   32 mg/ml
1 ml   Milli-Q water (Millipore)

- Weigh putrescine and dissolve in Milli-Q water.
- Store in 100 μl aliquots at −20 °C.

15. Sodium selenite (0.3 mM)

0.5 mg   Sodium selenite (MW = 172.9)   0.05 mg/ml
10 ml   Milli-Q water (Millipore)

- Weigh sodium selenite and dissolve in Milli-Q water.
- Store in 50 μl aliquots at −20 °C.

16. RO-20-1724 (35 mM)

9.7 mg   RO-20-1724 (MW = 278.4)   9.7 mg/ml
1 ml   DMSO

- Weigh RO-20-1724 and dissolve in DMSO.
- Sterile filtration in a laminar flow cabinet.
- Store in 100 μl aliquots at −20 °C.

17. CPT-cAMP (25 mM)

12.3 mg   CPT-cAMP (MW = 493.8)   12.3 mg/ml
1 ml   Milli-Q water (Millipore)

- Weigh CPT-cAMP and dissolve in Milli-Q water.
- Sterile filtration in a laminar flow cabinet.
- Store in 150 μl aliquots at $-20\,^\circ$C.

18. Collagen solution from pure substance (0.3 mg/ml)

30 mg   Collagen type VII
100 ml   Acetic acid (0.1% v/v)

- Sterile filtration of acetic acid.
- Dissolve collagen in acetic acid (under aseptic conditions) by stirring at 4 °C for 48 h.
- Centrifuge in a 50 ml tube for 2 h at 5000 rpm to seperate undissolved fibers from dissolved collagen.
- Store the collagen solution in a sterile bottle in the refrigerator.

19. Fibronectin solution (1 mg/ml, stock solution)

5 mg   Fibronectin, lyophilized, sterile
5 ml   Sterile filtered Milli-Q water (Millipore)

- Store in 1 ml aliquots at $-20\,^\circ$C, in a 100 ml bottle.

20. Poly-D-lysine (1 mg/ml, stock solution)

5 mg   Poly-D-lysine   1.0 mg/ml
5 ml   Milli-Q water (Millipore)

- Store in 1 ml aliquots at $-20\,^\circ$C, in a 100 ml bottle.

Isolation of Endothelial Cells from Fresh Calf Brain

Isolation procedure

1. Acquire fresh calf or bovine brain from the slaughterhouse. Store on ice and cover with 1 l PBS buffer while transporting to the laboratory. Do not use any brains that are bruised or mangled. Ideally the meninges should be intact, the tissue should be soft and have a pink hue.
2. Remove both cerebellum and brain stem, and then separate the hemispheres (use gloves). Transfer the hemispheres to a large petri dish with PBS at room temperature (RT). Use PBS (RT) to wash both hemispheres several times. Aspirate between washes.
3. Carefully and thoroughly remove the meninges, with your hands (sterile gloves). Work methodically starting on the anterior median edge (where the two lobes meet) over to the lateral edge. The two ends are often the most

difficult to peel. Note the areas where the meninges will not come off or where blood vessels have burst and be sure to stay away from these when collecting grey matter.

4. Wash the hemisphere carefully with PBS, also between the folders. At this point the grey matter should be pinkinsh grey in colour. Do not use a brain that has started to discolour and firm up.

5. Transfer the hemisphere to a clean 15 cm petridish containing DMEM + 10% FCS (RT). Change gloves (sterile).

6. Use a sterile razor blade to slice off pieces of the grey matter, leaving the white matter behind. Keep the brain moist by repeatedly bathing with medium. Transfer the pieces into a Falcon tube (225 ml) containing ($\pm$ 100 ml) DMEM + 10% FCS (RT).

7. Homogenize small aliquots of tissue in a 40 ml Wheaton homogenizer, first with pestle B and then pestle A. Fill the homogenizer 1/3–1/2 with brain pieces and add DMEM + 10% FCS (RT) to fill it up. Homogenize with pestle B until all large pieces are disrupted (8–10 strokes). Continue with pestle A until a homogenous suspension is obtained (8–10 strokes. Note the change in viscosity). Repeat this step until all brain tissue is homogenized. *Note*: Capillaries stick to plastic, therefore always flush any plastic ware (including pipettes) with medium first.

8. Filter homogenate through a 150 μm mesh, 30–40 ml per filter (mix well each time), using a 20 ml syringe and a Millipore filterholder. Rinse with 10 ml DMEM + 10% FCS (RT). Transfer the mesh with capillaries to a petri dish containing 10 ml DMEM + 10% FCS (RT) and flush the material off the mesh with the pipette. Transfer the suspension to 50 ml Falcon tubes and keep on ice. Use a fresh (autoclaved) piece of 150 μm mesh for the next aliquot. Discard the filtrate.
   Thaw the digest mix.

9. Spin the tubes for 5 min at 1000 rpm.

10. Aspirate the supernatants and remove the white top layer of the pellets with a pasteur pipette. Resuspend and pool the pellets in 20 ml (37 °C) digest mix. Incubate for 1 h at 37 °C (shake once or twice).

11. Stop the digestion by diluting 1:1 with DMEM + 10% FCS (to a volume of 50 ml).

12. Spin for 5 min at 1500 rpm.

13. Resuspend in 10 ml 'freeze mix' and put 1 ml into each cryovials.

14. Freeze and store the vials in the −80 °C freezer for a maximum of two weeks.

Plating of BMEC

Coat the culture flasks before use.

1. Thaw one vial with brain capillaries per 2 coated culture flasks (T75) in a 37 °C waterbath. Wipe the outside of the vial with 70% ethanol. In order to get rid of the (too) large capillaries, an extra filtration procedure is introduced.

2. Transfer the content of one vial to 10 ml DMEM + 10% FCS (flush the vial). Filter the suspension through a 200 μm mesh, using a 20 ml syringe and a

millipore filterholder. Rinse with DMEM + 10% FCS to a total volume of 50 ml and spin the filtrate for 5 min at 1000 rpm.

3. Transfer the content of one vial to 10 ml DMEM + 10% FCS (flush the vial) and spin for 5 min at 1000 rpm.
4. Aspirate the supernatant and resuspend in 20 ml DMEM + 10% FCS, 37 °C.
5. Gently pipette capillaries into culture flasks.
6. Leave in the incubator (37 °C, 10% CO_2) for 24 h for capillaries to attach, then change to growth medium.
7. Feed the cells every other day with 10 ml growth medium and passage between day 5 and 7 after palting, depending upon the speed of outgrowth.

Alternatively, for steps 3–5, plate straight into growth medium and change the medium the next day.

Coating of the culture flasks and filters for the BMEC

1. Dilute the collagen stock solution 30 times (from 0.3 mg/ml to 10 μg/ml) with 0.1% v/v acetic acid (sterile flitered, 0.2 μm) or use already diluted solution.
2. Add 6 ml collagen solution (10 μg/ml) per culture flask (T75) or 100 μl per filter (Costar 3413) for 2 h at room temperature.
3. Rinse the flasks or filters 3 times with PBS buffer before plating/seeding the cells.

For the culture flasks:

4. Dilute the fibrinectin stock solution 100 times (from 1 mg/ml to 10 μg/ml) with Milli-Q water (sterile filtered, 0.2 μm) or use already diluted solution.
5. Add 3 ml fibronectin (10 μg/ml in PBS, pH 7.4) for at least 30 min, aspirate and put (very carefully) on PBS buffer.
6. Use instantly.

Trypsinization procedure for BMEC

1. The bovine cells can be trypsinized 1 to a maximum of 2 weeks after isolation.
2. Before use coat the polycarbonate filters (Costar 3413, 6.5 μm, 0.4 μm) with 0.01 mg/ml collagen solution following the same procedure as for growth flasks (100 μl/filter). Be sure to rinse the whole filter.
3. All solutions used at 37 °C.
4. Wash the first growth flask two times with 10 ml PBS buffer.
5. Add 5 ml of trypsin-EDTA solution to the upstanding flask. Lay down the flask, under the microscope, and start the time. Tap the flask until most of the endothelial cells detach, then put the flask in upstanding position. Add 5 ml DMEM + 10% FCS to stop the trypsinization. **Maximal trypsinization time is 90 sec.**

6. Pour the cell suspension into a 50 ml Falcon tube.
7. Rinse the growth flask with 10 ml DMEM + 10% FCS and pour it in the Falcon tube as well. (Check if the cells are removed from the flask.)
8. Repeat steps 4 to 7 for the other growth flasks.
9. Pool the cell suspensions together in 50 ml Falcon tubes.
10. Spin for 5 min at 1500 rpm.
11. Resuspend and pool the pellets in 2 ml N2 medium.
12. Mix 12 μl of cell suspension with 12 μl crystalviolet (in duplicate) and count the cells.
13. Dilute the cell suspension with N2 + 50% ACM to the desired cell concentration. (100.000 cells/filter, for 0.3 cm^2 filters).
    (13a. Two days after seeding; change the medium with N2/ACM + cAMP. Incubate the cells overnight and use for experiment.)
14. Incubate the cells at 37 °C, 5% CO_2.
15. The maximum TEER is to be meassured 2–3 days after seeding.

Passaging of BMEC

Before use coat the polycarbonate filters (Costar 3413, 6.5 μm, 0.4 μm) and put them in the incubator with 750 μl growth medium (basolateral) and 150 μl (apical).

1. Rinse the cells with PBS (10 ml, 37 °C), and then with Versene (4 ml, 37 °C, Gibco).
2. Add 4 ml trypsin (37 °C, Sigma Endothelial Cell Trypsin), then observe continuously under the microscope.
   The endothelial cells are more sensitive to the trypsin than the contaminating pericytes, and should therefore come off first. The flask can be shaken gently to encourage the cells to come off *but* severe shaking and tapping of the flask on the beach will just result in the cells coming off in sheets taking the pericytes with them. When the majority of the endothelial cells have come off (but the pericytes are still stuck down, max. 1–2 min), then:
3. Wash the contents of the flask with at least 10 ml DMEM + 10% FCS (37 °C) to stop the trypsinization.
4. Repeat steps 1 to 3 for other flasks.
5. Centrifuge for 5 min at 1000 rpm and resuspend in 1 ml growth medium (37 °C) and then count the cells (one cell = 1; one cell cluster = 1).
6. Dilute the cell suspension with growth medium (37 °C) to the desired cell concentration (for example: 20,000 cells per 100 μl).
7. Seed the diluted cell suspension (100 μl) onto each filter.
8. Feed the cells every other day with 1 ml growth medium (37 °C). Remove the medium but leave approximately 50 μl behind on the cells, and then add the 1 ml to the apical side (the excess will flow to the basolateral side).
9. After 4 days, change the medium with N2/ACM (+ cAMP/RO-20-1724, if required, to increase the TEER).
10. Incubate the cells at 37 °C, 10% CO_2, overnight and use for the experiment.

III.2. A CO-CULTURE OF BRAIN CAPILLARY ENDOTHELIAL CELLS AND ASTROCYTES: AN *IN VITRO* BLOOD-BRAIN BARRIER FOR STUDYING DRUG TRANSPORT TO THE BRAIN

B. DEHOUCK, R. CECCHELLI AND M.-P. DEHOUCK

U 325 INSERM, Institut Pasteur, 1 rue Calmette, 59019 Lille,
Université de Lille I, Villeneuve d'Ascq, France

Investigations of the functional characteristics of brain capillaries have been facilitated by the use of cultured brain endothelial cells, and several groups are working towards a good *in vitro* model of the blood-brain barrier. The trick of our isolation procedure is to avoid the use of enzymes in the capillary separation procedure. The method of selection of cloned endothelial cells emerging from identified capillary avoids any contamination by pericytes and also any contamination by endothelial cells of larger vessel origin that could form relatively leaky areas in the monolayers. These cells are subcultured, at the split ratio of 1:20 (20-fold increase in the cultured surface), with no apparent changes in cell morphology. Retention of endothelial-specific characteristics is shown for brain capillary-derived endothelial cells up to passage 10, even after a frozen storage at passage 3.

We have subsequently developed a process of co-culture that closely mimics the *in vivo* situation by culturing brain capillary endothelial cells on one side of a filter and astrocytes on the underlying plastic well. Under these conditions, endothelial cells retain all the endothelial cell markers and the characteristics of the blood-brain barrier, including tight junctions and gamma-glutamyl transpeptidase. The transport of a number of labelled compounds through the monolayer was compared with their transport through the blood-brain barrier *in vivo*. There was a high correlation between the *in vivo* and *in vitro* permeability values for the tested substances, showing that the tightness of the monolayer is comparable with the *in vivo* properties.

INTRODUCTION

Several procedures have been developed in different laboratories to isolate and generate microvessel endothelial cells. In general, these procedures used either mechanical means or enzymatic digestion to disperse the brain tissue and obtain fragments of microvessels.

Using mechanical means, we developed a procedure of isolation that enabled us to obtain a preparation enriched with capillaries (Méresse *et al.*, 1989). Mechanical homogenization and filtration of cerebral microvessels were performed using a slightly modified method of Brendel *et al.* (1974). To obtain endothelial cells for culturing, capillaries separated from the great majority of arterioles and venules, were seeded, without a prior use of collagenase digestion, into extracellular matrix-coated dishes in accordance with the method of Gospodarowicz (1984a). Endothelial cell islands emerging from identified capillaries underwent a microtrypsinisation for amplification as described by Méresse *et al.* (1989). To mimic the *in vivo* situation, a process of co-culture was developed by culturing endothelial cells and astrocytes on opposite sides of a filter (Dehouck *et al.*, 1990a). To assess this co-culture model, the transport of different compounds through this *in vitro* model and the *in vivo* Oldendorf method was compared (Dehouck *et al.*, 1992a).

MATERIALS AND METHODS FOR SYSTEM ESTABLISHMENT

Chemicals

Crystalline bovine serum albumin was obtained from ICN. Gelatin and dextran T40 were from Sigma. Dulbecco's modified Eagle's medium (DMEM) was obtained from Gibco. Serums were from Hyclone laboratories (Logan, UT). Gentamycin and trypsin were purchased from Seromed. Tissue culture dishes were from Falcon Plastic. Bovine pituitary basic fibroblast growth factor (bFGF) was purified in our laboratory, using heparin-Sepharose affinity chromatography as previously described by Gospodarowicz *et al.* (1984b). Growth factor homogeneity was assessed by sodium dodecyl sulfate-polyacrylamide gel electrophoresis and its mitogenic activity was determined by its ability to stimulate proliferation of bovine aortic endothelial cells (Gospodarowicz *et al.*, 1976). Culture plate inserts (Millicell-CM 0.4 μm; 30 mm diameter) were obtained from Millipore.

Preparation of Extracellular Matrix-coated Dishes

Eyes from freshly slaughtered cows were obtained from a local slaughterhouse. The corneas were first washed with 95% ethanol. The cornea was then punctured near its edge (at the junction of the cornea and the sclera) with an 18-gauge needle. Dissecting scissors were inserted into the hole and the cornea was dissected out. It was then placed inverted (endothelial side up) in a 10 cm tissue culture dish. The cornea was carefully washed with phosphate-buffered saline (PBS). Once the corneal endothelium had been extensively washed it was delicately scraped with a groove director. The groove director was then dipped into a 3.5 cm tissue culture dish containing 2 ml of Dulbecco's modified Eagle's medium (DMEM) supplemented with 5% foetal calf serum (FCS), 5% calf serum (CS), 50 μg/ml gentamicin, and 2.5 μg/ml fungizone.

The plates were then incubated at 37 °C in a CO_2 incubator with 99% humidity and left undisturbed for 5 days, except for occasional examination under phase-contrast microscopy. After 5–6 days the media were changed, fresh media and basic fibroblast growth factor (bFGF) (1 ng/ml) were added every other day.

By day 10, well-developed colonies were visible. The primary cultures were trypsinised and passaged into one or two 6 cm-gelatinized dishes. bFGF was added every other day until the plates were nearly confluent. At that time, cells were passaged into ten to fifteen 6 cm dishes and grew in the presence of DMEM supplemented with 5% CS, 5% foetal CS, 5% dextran T40, and bFGF (1 ng/ml) added every other day. Once cultures had been confluent for 7 days, the medium was removed and the cultures were washed once with distilled water. To lyse and solubilize the cell monolayer, the cultures were then exposed to 20 mM NH_4OH in distilled water for 5 min. Final denudation of extracellular matrix (ECM) was obtained by extensive washing with distilled water. ECM-coated dishes were kept under sterile conditions in PBS for 2 months at 4 °C.

Preparation of Gelatin-coated Dishes

To coat the dishes with gelatin (Gospodarowicz and Massoglia, 1982), 2 ml of solution containing 0.2% gelatin in PBS were added to tissue culture dishes, which were left overnight at 4 °C. The next day, the solution was removed and the dishes were filled with culture medium just before plating the cells.

Mechanical Procedures for Initiation of Bovine Brain Capillary Endothelial Cells

Bovine brain was removed from freshly slaughtered animals. The brain cortex was cut into 2 mm^3 fragments and washed twice with PBS. Tissue fragments were resuspended in 2 volumes of PBS and homogenized by 15 up-and-down strokes in a 40 ml glass homogenizer fitted with a glass pestle (0.152 mm clearance). The homogenate was passed through a 180 μm nylon sieve, and the filtrate, containing microvessels, was rehomogenized with a second glass pestle (0.076 mm clearance). Finally, the capillaries were collected on a 60 μm nylon sieve, washed in PBS and resuspended in DMEM supplemented with 10% CS, 10% horse serum (HS), 50 μg/ml gentamycin, and 2.5 μg/ml fungizone. The freshly isolated microvessel suspension was found to consist of capillaries when monitored under light and phase-contrast microscopy.

The microvessel preparation was seeded immediately at a density of 5–10 fragments/cm^2 into ECM-coated dishes containing DMEM supplemented with 10% CS and 10% HS. After 2 h incubation in a humidified incubator (37 °C) in an environment of 5% CO_2/95% air, most capillaries had adhered to ECM.

The medium was changed in order to remove nonadherent microvessels. Four to five days after seeding the first endothelial cells migrated out from the capillaries and began to form microcolonies. The medium was changed and bFGF (1 ng/ml) was added every other day. When the endothelial cell colonies were sufficiently large for cloning (50–100 cells) (Figure 1A), the medium was removed and dishes were washed twice with calcium and magnesium-free PBS (PBS-CMF). Using a glass Pasteur pipette connected to a vacuum, each endothelial cell colony was isolated by making a dry ring by means of peripheral mechanical suction and then covered with 10–20 μl of PBS-CMF solution containing 0.05% trypsin and 0.02% EDTA. The trypsinized BBCE cells were harvested using a pipette equipped with a 200 μl pipette tip, seeded on 3.5 cm gelatin-coated dishes in the presence of DMEM supplemented with 20% CS and bFGF (1 ng/ml added every other day) and incubated in a humidified incubator (37 °C) in an environment of 5% CO_2/95%.

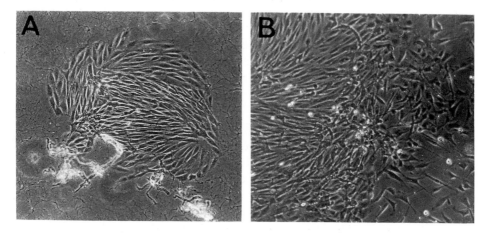

Figure 1 Pure endothelial cell island in primary culture.

BBCE Cell Culture Conditions

After cloning (first passage), endothelial cells were harvested as described below and seeded on a 6 cm gelatin-coated dish. After 6–8 days, confluent cells were subcultured at a split ratio of 1:20 (20-fold increase in the cultured surface). Cells at the third passage were stored in liquid nitrogen for several months. For freezing, cells of a 6 cm dish were harvested as for subculturing, centrifuged at 600 g for 5 min, resuspensed in growth medium containing 10% dimethyl sulphoxide (vol/vol) and 10 mM HEPES, pH 7.2 cooled for 1 h at 4 °C and subsequently for 4 h at −70 °C, and stored in liquid nitrogen. For experiments cells were rapidly thawed at 37 °C and seeded onto three 6 cm gelatin-coated dishes. Once they had reached confluence, cells were subcultured up to passage 8.

For subculturing, confluent stock dishes of BBCE cells grown in the presence of DMEM supplemented with 10% CS, 10% HS, gentamycin, fungizone, and bFGF were washed twice with PBS-CMF and then exposed to a PBS-CMF solution containing 0.05% trypsin and 0.02% EDTA. When cells became round, they were suspended in their growth medium and the cells were seeded on gelatin coated dishes at a split ratio of 1:20.

Co-culture of Brain Capillary Endothelial Cells and Astrocytes

Rat astrocytes

Primary cultures of mixed astrocytes were made from cerebral cortex of newborn rat. After the meninges had been removed, the brain tissue was gently forced through a 82 μm nylon sieve, as described by Booher and Sensenbrenner (1972). Astrocytes were plated on 6 multiwell dishes at a concentration of 1.2×10^5 cells/ml in 2 ml of DMEM supplemented with 10% FCS and the medium was changed twice a week. Three weeks after seeding, cultures of astrocytes stabilized and were used for experiments. The astrocytes were characterized with glial fibrillary acidic protein (GFAP), and more than 95% of the population was GFAP positive (Dehouck et al., 1990b).

Co-culture of endothelial cells and astrocytes

Culture plate inserts were coated on the upperside with rat tail collagen prepared by a modification of the method of Bornstein (1958). Cultures of astrocytes were prepared as described above. After three weeks, coated filters were set in 6 multiwell dishes containing astrocytes and endothelial cells were plated on the upperside of the filters in 1.5 ml of medium with a concentration of 4×10^5 cells/ml. The medium used for the co-culture was DMEM supplemented with 10% CS, 10% HS, 2 mM glutamine, 50 μg/ml gentamycin and 1 g/ml bFGF. This medium was changed every other day. Under these conditions, endothelial cells form a confluent monolayer after 7 days. Experiments were performed 5 days after confluence. This arrangement readily permits the use of different cell types, which were separated easily after co-culture by removing the insert.

Specific Applications: Transendothelial Transport Studies

On the day of the experiments, Ringer-HEPES (150 mM NaCl, 5.2 mM KCl, 2.2 mM $CaCl_2$, 0.2 mM $MgCl_2$, 6 mM $NaHCO_3$, 2.8 mM glucose, 5 mM HEPES) was added to the lower compartments of a six-well plate (2 ml per well). One filter containing a confluent monolayer of bovine brain capillary endothelial cells was transferred into the first well of the six-well plate containing Ringer-HEPES. Ringer-HEPES containing ^{14}C-labelled drugs or ^{3}H-labelled drugs (2 ml) was placed in the upper compartment. At different times after addition of the labelled compound, the filter was transferred to another well of the six-well plate to minimize the possible passage from the lower to the upper compartment as described in Figure 2. Triplicate monolayers and triplicate filters coated only with collagen were assayed for each drug. An aliquot from each lower compartment and 20 μl from the initial solution containing the labelled drug was placed in a scintillation vial, and the radioactivity was determined. The same experiment was also carried out using labelled drugs. In these cases, quantities of the drug in the lower compartments were measured by HPLC.

Permeability calculation were performed as described by Siflinger-Birnboim et al. (1987). To obtain a concentration-independent transport parameter, the clearance principle was used. The increment in cleared volume between successive sampling events was calculated by dividing the amount of solute transport during the interval by the donor chamber concentration. The total cleared volume at each time point

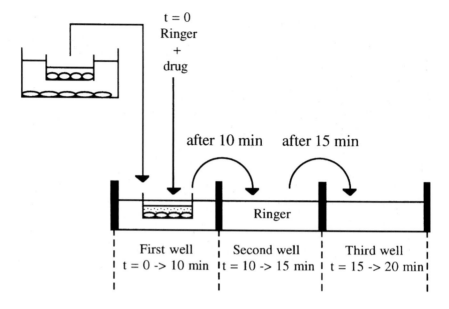

Figure 2 Transendothelial transport study.

was calculated by summing the incremental cleared volumes up to the given time point:

$$\text{clearance (ml)} = \text{Cl (ml)} = \frac{X \text{ (dpm)}}{Cd \text{ (dpm/ml)}}$$

where X (dpm) is the amount of drug in the receptor chamber and Cd (dpm/ml) is the donor chamber concentration at each time-point.

During the 45-min experiment, the clearance volume increased linearly with time. The average volume cleared was plotted versus time, and the slope was estimated by linear regression analysis. The slope of the clearance curves for the culture was denoted PSt, where PS was the permeability × surface area product (in ml per min). The slope of the clearance curve with the control filter coated only with collagen was denoted PSf.

The PS value for the endothelial monolayer (PSe) was calculated from

$$\frac{1}{PSe} = \frac{1}{PSt} - \frac{1}{PSf}$$

The PSe values were divided by the surface area of the Millicell-CM (4.2 cm^2) to generate the endothelial permeability coefficient (Pe, in cm per min).

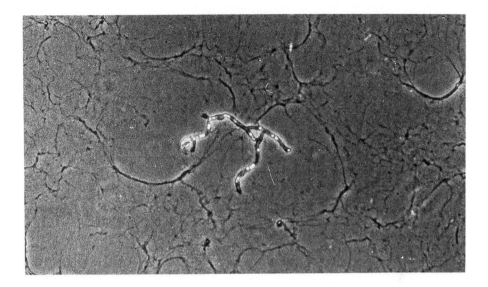

Figure 3 Capillary adhering to extracellular matrix-coated dishes.

PROBLEM SOURCES AND QUALITY CONTROL

Disruption of brain tissue by mechanical dispersion and filtration techniques, without enzymatic treatment enables us to isolate a microvascular network consisting predominantly of capillaries with some arterioles and venules. Only capillaries adhere on ECM-coated dishes (Figure 3).

Only pure endothelial cell islets emerging from identified capillaries, such as the one shown in Figure 1A, are cloned and subcultured. When using this technique, pericyte contamination is avoided (Figure 1B). Confluent brain capillary endothelial cells form a monolayer of small tightly packed, non-overlapping, contact inhibited cells (Figure 4). In culture, cells present the characteristics of vascular endothelial cells: factor VIII-related antigen, angiotensin converting enzyme, non-thrombogenic properties and production of PGI_2 (Méresse *et al.*, 1989). Typical properties of the cerebral endothelium are expressed: the presence of tight junctions and monoamine oxidase activity (Méresse *et al.*, 1989). These cells, absolutely free of pericytes can be passaged repeatedly up to 10 passages at a split ratio of 1:20, without the loss of endothelial cell markers.

Endothelial cells *in solo* culture, however, lose gamma-glutamyl transpeptidase activity when compared with isolated microvessels. By culturing endothelial cells and astrocytes on opposite sides of a filter, gamma-glutamyl transpeptidase is reinduced in endothelial cells (Dehouck *et al.*, 1990a). Moreover, the permeability of the endothelial monolayer for inulin is decreased twice when the endothelial cells are co-cultured with astrocytes (Dehouck *et al.*, 1992b). To assess the integrity of the monolayers, transfer of sucrose and inulin (substances that diffuse very slowly in physiological conditions) is performed as described in Figure 2, before each experiment. This can be done quickly and is more reproducible than the measurement of the transendothelial electrical resistance (TEER; see also chapters II.2 and III.1).

The blood-brain barrier permeability of our co-culture model is measured for 11 radiolabelled compounds and compared with the blood-brain barrier permeability *in vivo* (Dehouck *et al.*, 1992a). The *in vivo* and *in vitro* permeability values show a strong correlation. Moreover, the *in vitro* permeability of compounds that cross the blood-brain barrier via carrier mediation is in the same range as that of the blood-brain barrier permeability *in vivo*, indicating that in our model the specific transporters are always present.

CONCLUDING REMARKS

Primary brain microvessels are, at least to start with, a mixture of cells of capillary, arteriolar, and venular origin since the microvessel fraction obtained from the brain consists mainly of capillaries as well as arterioles and venules. An obvious advantage of the use of cloned endothelial cells emerging from identified capillaries is that the culture is not contaminated by endothelial cells of arteriolar and venular origin. Moreover in primary culture, pure endothelial cell islets (Figure 1A) and pure pericyte islands (Figure 5) can be micro-trypsinised separately.

The subculture technique enabled us to circumvent the culture limitations of primary cultures and to provide a large quantity of monolayers in a short time as explained in Figure 6.

The relative ease with which such co-cultures can be produced in large quantities (hence mass production and eventually, cryopreservation) offers advantages, including rapid assessment of the potential permeability and metabolism of a drug, and the opportunity to elucidate the molecular transport mechanism of substances across the blood-brain barrier.

REFERENCES

Booher, J. and Sensenbrenner, M. (1972) Growth and cultivation of dissociated neurons and glial cells from embryonic chick, rat and human brain in flask culture. *Neurobiology*, **2**, 97–105.

Bornstein, M.B. (1958) Reconstituted rat tail collagen used as a substrate for time tissue cultures on coverslips in maximow slides and roller tubes. *Lab. Invest.*, **7**, 134–139.

Brendel, K., Meezan, E. and Carlson, E.C. (1974) Isolated brain microvessels: A purified metabolically active preparation from bovine cerebral cortex. *Science*, **185**, 953–955.

Dehouck, M.P., Méresse, S., Delorme, P., Fruchart, J.C. and Cecchelli, R. (1990a) An easier, reproducible, and mass-production method to study the blood-brain barrier *in vitro*. *J. Neurochem.*, **54**, 1798–1801.

Dehouck, M.P., Méresse, S., Delorme, P., Torpier, G., Fruchart, J.C. and Cecchelli, R. (1990b) The blood-brain barrier *in vitro*: Co-culture of brain capillary endothelial cells and astrocytes. *Circulation et metabolisme du Cerveau*, **7**, 151–162.

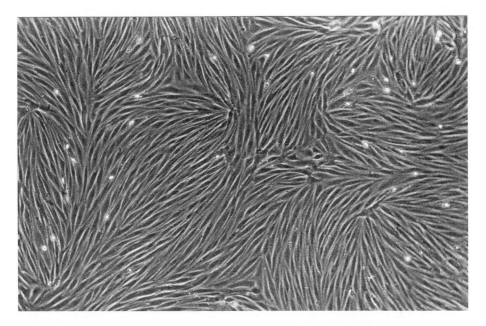

Figure 4 Confluent monolayers of brain capillary endothelial cells in their third passage.

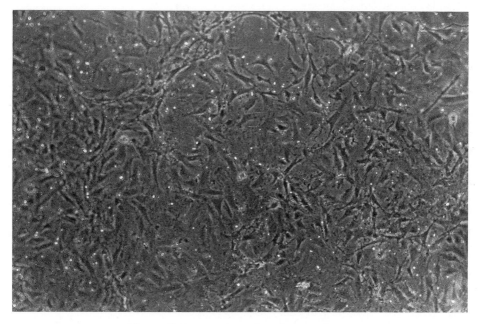

Figure 5 Pure pericytes in primary culture.

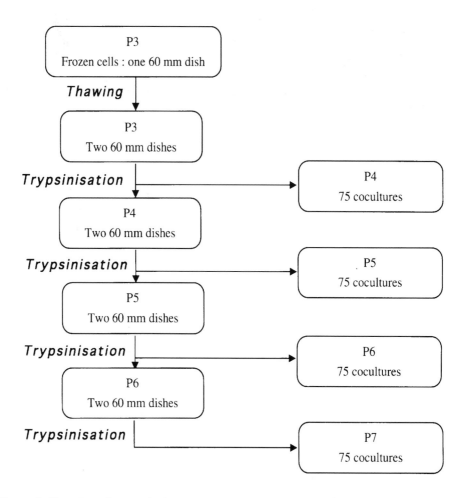

Figure 6 Use of capillary endothelial cells after thawing. After P7, if the monolayer looks good under contrast microscopy, an increase in sucrose and inulin permeability is observed.

Dehouck, M.P., Jolliet-Riant, P., Brée, F., Fruchart, J.C., Cecchelli, R. and Tillement, J.P. (1992a) Drug transfer across the blood-brain barrier: Correlation between *in vitro* and *in vivo* models. *J. Neurochem.*, **58**, 1790–1797.

Dehouck, M.P., Méresse, S., Dehouck, B., Fruchart, J.C. and Cecchelli, R. (1992b) *In vitro* reconstituted blood-brain barrier. *Journal of Controlled Release*, **21**, 81–92.

Gospodarowicz, D., Moran, J., Braun, D. and Birdwell, C. (1976) Clonal growth of bovine vascular endothelial cells: Fibroblast growth factor as a survival agent. *Proc. Natl. Acad. Sci. USA*, **73**, 4120–4124.

Gospodarowicz, D. and Massoglia, S.L. (1982) Plasma factors involved in the *in vitro* control of proliferation of bovine lens cells grown in defined medium: Effect of fibroblast growth factor on cell longevity. *Exp. Eye Res.*, **35**, 259–270.

Gospodarowicz, D. (1984a) Preparation of extracellular matrices produced by cultured bovine corneal endothelial cells and PFHR-9 endodermal cells, their use in cell culture. In: *Methods in Molecular and Cell Biology*, **1**, 275–293. New York: Alan R. Liss, Inc.

Gospodarowicz, D., Cheng, J., Lui, G.M., Baird, A. and Böhlen, P. (1984) Isolation of brain fibroblast growth factor by heparin Sepharose affinity chromatography: Identity with pituitary fibroblast growth factor. *Proc. Natl. Acad. Sci. USA*, **81**, 6963–6967.

Méresse, S., Dehouck, M.P., Delorme, P., Bensaïd, M., Tauber, J.P., Delbart, C., Fruchart, J.C. and Cecchelli, R. (1989) Bovine brain endothelial cells express tight junctions and monoamine oxydase activity in long-term culture. *J. Neurochem.*, **53**, 1363–1371.

Siflinger-Birnboim, A., Del Becchio, P.J., Cooper, J.A., Blumenstock, F.A., Shepard, J.N. and Malik, A.B. (1987) Molecular sieving characteristics of the cultured endothelial monolayer. *J. Cell. Physiol.*, **132**, 111–117.

III.3 PRIMARY CULTURES OF BOVINE BRAIN MICROVESSEL ENDOTHELIAL CELLS

D. LECHARDEUR, B. SCHWARTZ AND D. SCHERMAN

UMR 133 CNRS/Rhône-Poulenc Rorer, Centre de recherche de Vitry-Alfortville, Bâtiment Monod, 13, quai Jules Guesde, BP 14, 94403 Vitry/Seine, France

The use of enzymatic procedures for the isolation of bovine brain microvessels allows to obtain relatively pure primary cultures of endothelial cells. These cells are useful for biochemical and cellular analyses *in vitro*. They are also promising for obtaining a pharmacologically relevant *in vitro* model of the blood-brain barrier (BBB), which can be used for a preliminary selection of the brain penetration of drug candidates.

Brain microvessel endothelial cells might also be the starting material for obtaining an immortalized clonal cell line. This is described in chapter II.3 for the rat.

INTRODUCTION

Several groups have attempted to use *in vitro* cellular models of blood-brain-barrier (BBB) (Rubin *et al.*, 1991; Dehouck *et al.*, 1992) as a convenient and versatile method to study drug transfer across the BBB, e.g. in initial prediction of penetration of novel drugs into the brain. The prerequisite *in vivo* characteristics that these models must possess are described in chapter II.3.

This chapter (and chapter II.3 for the rat system) aim to illustrate the techniques and strategies used to obtain proper primary cultures. The cell biology techniques or materials used are described in detail in Lechardeur and Scherman (1995), and Lechardeur *et al.* (1995).

PRIMARY CULTURE OF BOVINE BRAIN CAPILLARY ENDOTHELIAL CELLS

Bovine endothelial cell preparations were performed according to the method of Audus and Borchardt (1986). Bovine brains were collected in a slaughterhouse and transported in cold HBSS containing 100 U/ml penicillin, 100 μg/ml streptomycin and 100 U/ml mycostatin. Pial vessels and meninges were dissected out, grey matter of cortices minced into small fragments and incubated for 3 h at 37 °C under agitation in the preparation medium (DMEM 4.5 g/l glucose, 25 mM HEPES, 100 U/ml penicillin, 100 μg/ml streptomycin) containing 10 μg/ml DNAse 1 from bovine pancreas and 1 mg/ml collagenase/dispase (Boehringer).

After centrifugation at 1000 g for 10 min, the pellet was resuspended in a solution of BSA solubilized in the preparation medium (1/3 pellet volume and 2/3 BSA solution volume, 25% BSA final concentration) and centrifuged at 2500 g for 10 min.

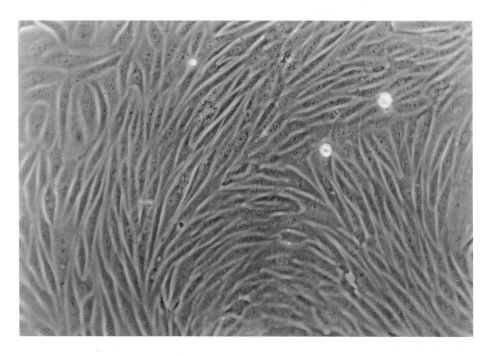

Figure 1 Bovine brain capillary endothelial cells in primary culture. Spindle-shaped endothelial cells obtained as described constitute, at confluency, a cellular monolayer stable for several days. Phase contrast microscopy (×100).

Fat, cell debris and myelin floating on the top of the BSA solution were discarded and the pellet containing isolated microvessels was resuspended in the preparation medium containing 10 μg/ml DNAse 1, and 1 mg/ml collagenase/dispase for an incubation time of 5 h at 37 °C or 0.5 mg/ml for an overnight incubation time at 37 °C. Microvessels were then separated from cell debris and erythrocytes by centrifugation at 2000 g for 10 min on a pre-established gradient of 50% Percoll/50% DMEM. Capillaries were collected in the intermediate part of the gradient, washed with preparation medium and resuspended in culture medium consisting of DMEM, 4.5 g/l glucose supplemented with 10% horse serum (Eurobio), 100 U/ml penicillin, 100 μg/ml streptomycin, 2 mM L-glutamine and 5 ng/ml bFGF(Boehringer) or 25 μg/ml endothelial cell growth supplement (SIGMA).

Microvessels were seeded at a density of about 10,000 capillaries/cm^2 onto collagen-coated plastic dishes. Coating was performed with a rat tail collagen (Sigma) solution in 0.1 N acetic acid prepared at 250 μg/ml, poured so as to cover the whole dish (about 250 μl/cm^2) and left for 2 h at 37 °C before being rinsed twice with PBS.

Primary cells showing a spindle shaped morphology, classical for capillary endothelial cells, were maintained in culture for several days (Figure 1) and were stable at confluency.

CONCLUDING REMARKS

Our experience has shown that bovine cells are easier to manipulate than rat cells; we obtained from bovine material the largest quantity of cells, with cultures that divided more quickly and with less numerous contaminating cells.

The use of primary cultures seems very limited by the presence of contaminating cells which did not allow us to obtain tight monolayers in dual chamber systems. This prompted us and others (Durieu-Trautmann *et al.*, 1991) to develop a clonal immortalized cell line, which might present several advantages as described in chapter II.3.

ACKNOWLEDGEMENTS

This work was sponsored by the CNRS, the Rhône Poulenc Rorer Society, and the BIOAVENIR program of French Ministry of Education and Research. We are indebted to Dr. Monique Santarromana for excellent help and suggestions in the realization of this commentary.

REFERENCES

Audus, K.L. and Borchardt, R.T. (1986) Bovine brain microvessels endothelial cell monolayers as a model system for the blood-brain barrier. *Ann. N. Y. Acad. Sci.*, **507**, 9–18.

Dehouck, M.P., Méresse, S., Dehouck, B., Fruchart, J.C. and Cecchelli, R. (1992) *In vitro* reconstituted blood-brain barrier. *J. Controlled Release*, **21**, 81–92.

Durieu-Trautmann, O., Foignant-Chaverot, N., Perdomo, J., Gounon, P., Strosberg, A.D. and Couraud, P.O. (1991) Immortalization of brain capillary endothelial cells with maintenance of structural characteristics of the brain barrier. *In Vitro Cell Dev. Biol.*, **27A**, 771–778.

Lechardeur, D., Schwartz, B., Paulin, D. and Scherman, D. (1995) Induction of blood-brain barrier differentiation in a rat brain derived endothelial cell line. *Exp. Cell Res.*, **220**, 161–170.

Lechardeur, D. and Scherman, D. (1995) Functional expression of the P-glycoprotein *mdr* in primary cultures of bovine capillary endothelial cells. *Cell Biol. Tox.*, **11**, 219–230.

Rubin, L.L., Hall, D.E., Porter, S., Barbu, K., Cannon, C., Horner, H.C., Janatpour, M., Liaw, C.W., Manning, K., Morales, J., Tanner, L.I., Tomaselli, K.J. and Bard, F. (1991) A cell culture model of the blood-brain barrier. *J. Cell Biol.*, **115**, 1725–1735.

IV.1. PREPARATION OF PRIMARY CULTURE FROM NEWBORN PIGS

M.A. DELI,[1] C.S. ÁBRAHÁM,[2] N.T.K. DUNG[1] AND F. JOÓ[1,†]

[1,†]*Laboratory of Molecular Neurobiology, Institute of Biophysics,*
Biological Research Center of the Hungarian Academy of Sciences,
P.O. Box 521, H-6701 Szeged, Hungary
[2]*Department of Paediatrics, Albert Szent-Györgyi Medical University,*
P.O. Box 471, H-6701 Szeged, Hungary

Primary cultures of cerebral endothelial cells were obtained from cortical grey matter of newborn pigs (*Sus scrofa*) using a two-step enzymatic digestion followed by a Percoll gradient centrifugation. Porcine cerebral endothelial cells (PCECs) formed a monolayer of spindle-shaped, tightly attached cells on collagen matrix. PCECs were specifically stained for Factor VIII, and bound *Bandeiraea simplicifolia* isolectin I-B$_4$ (BS-I-B$_4$). On average, a value of 136 $\Omega.cm^2$ for transendothelial electrical resistance (TEER) was measured for the monolayers. Primary cultures were used to study the direct effect of tumour necrosis factor-α (TNFα) on transendothelial permeability.

INTRODUCTION

Several procedures have been worked out for culturing PCECs in different laboratories, most of them have adapted protocols established to obtain rodent or bovine CECs.

Tontsch and Bauer (1989) isolated microvessels from porcine cortex by repeated homogenization and centrifugation steps, without sieving or gradient centrifugation. The endothelial cells were then separated from surrounding basement membrane by a brief and mild collagenase digestion. Robinson *et al.* (1990) used 4-6-month-old Yucatan minipigs, and separated PCECs by sieving, Ficoll-Paque gradient centrifugation, and a final incubation with collagenase. Our protocol, developed from the rat cerebral endothelial cell isolation procedure (see Chapter A.II.5), is the closest to the method of Galla *et al.* (Mischeck *et al.*, 1989; see Chapter A.IV.2). We used grey matter of newborn pig brains as a starting material, collagenase and collagenase-dispase for enzymic digestions, and finally PCECs clusters were separated by Percoll gradient centrifugation step before seeding.

MATERIALS AND METHODS

Animals

Newborn pigs, of either sex, weighing 1.08–1.62 kg, from a local cooperative farm were used in the experiments.

[†]Deceased.

Chemicals

See also chapter A.II.5 for product list.

Products	Supplier and catalogue No
Anaesthesia	
Ketamine hydrochloride (Ketanest®)	Parke-Davis, Morris Plains, NJ, USA
Pipecuronium bromatum (Arduan®)	Richter Gedeon Co., Budapest, HUN
Establishment of primary cultures	
Characterisation } *	

*See Chapter A.II.5.

Establishment of Primary Cultures

Dissection of brains

The animal was anaesthetised with ketamine hydrochloride (10 mg/kg body weight i.m.) and was given pipecuronium bromatum (0.2 ml/kg body weight i.v.) and potassium chloride (i.v.). After a thorough rinse with 70% ethanol, and then with iodine in 70% ethanol, the head was cut, and placed into a sterile glass dish. Forebrains were removed from the skull using sterile forceps and scissors, and put in ice-cold sterile phosphate buffered saline (PBS, without calcium and magnesium, pH 7.4). In the laminar flow box meninges were removed on sterile filter paper (Whatman 3M) and at the same time white matter was 'peeled off' with the aid of fine curved forceps. Grey matter was carefully collected from the filter paper (meninges tend to stick to it) and minced to approximately 1 mm³ pieces using sterile disposable scalpels in the first incubation medium (3 mg/ml collagenase CLS2 from Worthington, 1 mg/ml bovine serum albumin (BSA) in Dulbecco's modified Eagle's medium (DME) containing antibiotics) in a sterile glass petri dish.

Enzyme digestions

The minced tissue was transferred into 2 centrifuge tubes (35 ml, Oakridge-type with screw cap) with the rest of the collagenase solution (total: 30 ml/brain) and triturated with a pipette (10 up and down), and then incubated at 37°C for 1.5 h in a shaking waterbath. After this incubation, 15 ml of cold 25% BSA-DME were added to the homogenate in each tube, mixed well by trituration and centrifuged at 1000 g for 20 min. The myelin layer and the supernatant was aspirated from the tubes, the pellets were washed once in DME (1000 g for 10 min), then pooled, and further incubated in the waterbath for a maximum of 2 h in 15 ml of the second incubation medium containing 1 mg/ml collagenase-dispase in DME.

Percoll gradient centrifugation

The cell suspension was centrifuged (700 g for 5 min). The pellet was suspended in 2 ml DME and carefully layered onto a continuous 33% Percoll gradient and centrifuged at 1000 g for 10 min. For the gradient 10 ml Percoll, 18 ml PBS, 1 ml fetal calf serum (FCS) and 1 ml 10 × concentrated PBS were mixed, sterile filtered and centrifuged at 4°C, 30000 g for 1 h.

Plating and feeding cells

The band of the endothelial cell clusters (clearly visible as a white-greyish layer above the red blood cells) was aspirated and washed twice in DME (1000 g, 10 min). The cells were then suspended in culture medium (DME/F-12 containing 100 U/ml penicillin, 100 μg/ml streptomycin, 50 μg/ml gentamicin, 2 mM glutamine, 20% heat inactivated FCS) and seeded onto rat tail collagen-coated 35 mm plastic dishes or 25 mm cell culture inserts. The culture medium was changed on the next day, and later on, every third day.

Yield

We could obtain 150 cm^2 confluent primary culture of PCECs from 1 brain, approximately equivalent to 15–16 pieces of 35 mm tissue culture Petri dish.

Comments

The incubation media for enzymic digestions were always prepared freshly with lyophilised enzymes, and then sterilised by filtration. Their pH was adjusted to 7.4. During the separation we used only plasticware. If it was necessary to use any kind of glassware, we coated it before use with BSA.

SPECIFIC APPLICATION

The effect of recombinant human TNFα has been studied on the blood-brain barrier (BBB) permeability and pial vasoreactivity after an intracisternal injection in newborn pigs *in vivo* (Megyeri *et al.*, 1992).

For comparison with the results obtained on the *in vivo* system, the direct effects of the cytokine were assessed on primary cultures of PCECs. Viability of the cells was not changed significantly even after a 4 h incubation with 500 IU/ml TNFα. Cells were concomitantly loaded with fluorescein diacetate and ethidium bromide before treatment, and the result was evaluated using fluorescence microscopy (cytoplasm of living cells was green, nuclei of dead cells were red).

TEER of control and TNFα-treated (50–500 IU/ml) monolayers cultured on 25 mm inserts (Falcon) was measured using a WPI instrument. TNFα decreased TEER significantly in PCEC monolayers: $139.55 \pm 6.7 \ \Omega.cm^2$ in control ones $vs.$ 66.15 $\pm 4.6 \ \Omega.cm^2$ after treatment with 50 IU/ml TNFα for 2 h (n = 6, X $\pm$ S.D.). These results, which indicated an enhanced paracellular flux induced by TNFα, were in accordance with the observations on $in\ vivo$ models (Megyeri $et\ al.$, 1992; Temesvári $et\ al.$, 1995), where BBB permeability for small molecular weight substances was elevated.

PROBLEM SOURCES AND QUALITY CONTROL

PCECs formed a non-overlapping continuous monolayer at the end of the first week (Chapter A.II.5, Figure 1e). The morphology of the cells was 'fibroblast-like': fusiform cell-shape; oval, centrally positioned nuclei; neighbouring cells tightly apposed to each-other. PCECs gave specific immunohistochemical staining with anti-FVIII antibody, bound the galactose-specific BS-I-B$_4$ isolectin (for methods of characterisation see Chapter A.II.5). For PCEC monolayers grown on 25 mm inserts (Falcon), a value of 136 $\Omega.cm^2$ TEER was obtained in average.

In confluent primary cultures older than 7–10 days $in\ vitro$ angiogenesis, occasionally contamination by non-endothelial cells, could be observed. Unlike in the case of rat CECs, no methods for selective cytolysis of contaminating pericytes and/or astrocytes in PCEC cultures have been described in the literature yet.

CONCLUDING REMARKS

Studies on permeability and transport across PCEC monolayers ought to be carried on monolayers with TEER equal or higher than 500 $\Omega.cm^2$. Methods, which are successfully used to enhance the resistance of bovine cerebral endothelial cell monolayers, like co-culture of cerebral endothelial cells with astrocytes (Dehouck $et\ al.$, 1990), and/or treatment of monolayers with compounds that increase intracellular cyclic AMP concentration, thereby enhance the tightness of tight junctions between endothelial cells (Rubin $et\ al.$, 1991; Deli $et\ al.$, 1995) should be adapted to the porcine system, to create a more reliable $in\ vitro$ model of the BBB.

REFERENCES

Dehouck, M-P., Méresse, S., Delorme, P., Fruchart, J.C. and Cecchelli, R. (1990) An easier, reproducible and mass-production method to study the blood-brain barrier $in\ vitro$. $J.\ Neurochem.$, **57**, 1798–1801.

Deli, M.A., Dehouck, M-P., Ábrahám, C.S., Cecchelli, R. and Joó, F. (1995) Penetration of small molecular weight substances through cultured bovine brain capillary endothelial cell monolayers: the early effects of cyclic adenosine 3′,5′-monophosphate. $Exp.\ Physiol.$, **80**, 675–678.

Megyeri, P., Ábrahám, C.S., Temesvári, P., Kovács, J., Vas, T. and Speer, C.P. (1992) Recombinant human tumor necrosis factor α constricts pial arterioles and increases blood-brain barrier permeability in newborn piglets. *Neurosci. Lett.*, **148**, 137–140.

Mischeck, U., Meyer, J. and Galla H.-J. (1989) Characterization of γ-glutamyl transpeptidase activity of cultured endothelial cells from porcine brain capillaries. *Cell Tissue Res.*, **256**, 221–226.

Robinson, D.H., Kang, Y.-H., Deschner, S.H. and Nielsen, T.B. (1990) Morphologic plasticity and periodicity: Porcine cerebral microvascular cells in culture. *In vitro*, **26**, 169–180.

Rubin, L.L., Hall, D.E., Parter, S., Barbu, C., Cannon, C., Horner, H.C., Janatpour, M., Liaw, C.W., Manning, K., Morales, J., Tanner, L.I., Tomaselli, K.J. and Bard, F. (1991) A cell culture model of the blood-brain barrier. *J. Cell Biol.*, **115**, 1725–1735.

Temesvári, P., Joó, F., Kovács, J. and Ábrahám, C.S. (1995) Ischemia/reperfusion-induced alteration of blood-brain barrier transport in newborn pigs. *Am. J. Physiol.*, **269**, H750–H751.

Tontsch, U. and Bauer, H.-C. (1989) Isolation, characterization, and long-term cultivation of porcine and murine cerebral capillary endothelial cells. *Microvasc Res.*, **37**, 148–161.

IV.2. PREPARATION OF ENDOTHELIAL CELLS IN PRIMARY CULTURES OBTAINED FROM THE BRAINS OF 6-MONTH-OLD PIGS

B. TEWES, H. FRANKE, S. HELLWIG, D. HOHEISEL, S. DECKER,
D. GRIESCHE, T. TILLING, J. WEGENER AND H.-J. GALLA

Institut für Biochemie, Wilhelm-Klemm-Str.2 D-48149 Münster, Germany

Primary cultures of porcine brain capillary endothelial cells (BCEC) were obtained by a sequence of enzymatic digestion and centrifugation steps using the modified methods of Bowman *et al.* (1983) and Mischeck *et al.* (1989). Grown on permeable filter supports, the primary cultured porcine BCEC develop electrical resistance values up to 130 $\Omega \cdot cm^2$ and are suited for the study of transendothelial transport. Pericyte contaminations were avoided by subcultivating the cells.

INTRODUCTION

Primary cultures, passaged cells, cell lines and clones have been used for multiple applications on blood-brain barrier (BBB) research in the last two decades. However it was shown recently that BCEC *in vitro* lose their BBB-specific differentiation during proliferation in culture, as clearly demonstrated by a drastic decrease in alkaline phosphatase (ALP)- and gamma-glutamyl-transpeptidase (γ-GT) enzymatic activity. Therefore a model for BBB characteristics *in vitro* should be based on primary cultures. Referring to the common method of Bowman *et al.* (1983) we established a primary culture of porcine brain capillary endothelial cells (BCEC). As porcine brains are easily available in a sufficient amount and porcine physiology is closely related to human, porcine BCEC-cultures provide a most suitable *in vitro* model of the BBB.

MATERIALS AND METHODS FOR SYSTEM APPLICATION

Materials

Collagenase/Dispase	Sigma, Deisenhofen
Collagen G	Seromed, Berlin
Rat tail collagen	Boehringer, Mannheim
Dextran	Sigma, Deisenhofen
Dispase from *Bacillus polymyxa*	Boehringer, Mannheim
Earle's Medium M199	Seromed, Berlin
Earle's Buffer (10×) with Phenol red	Seromed, Berlin

EDTA	Sigma, Deisenhofen
Gentamicin (10 mg/ml)	Seromed, Berlin
Ox-serum	PAA, Linz (A)
Penicillin/Streptomycin (10.000 U/ml)	Seromed, Berlin
Percoll	Sigma, Deisenhofen
Cutter with rolling blades	Karstadt, Münster
Centrifuge tubes (250 ml)	Beckman, München
Centrifuge tubes (50 ml)	Greiner, Solingen
Culture dishes	Greiner, Solingen
Culture flasks	Nunc, Wiesbaden
Transwell filter inserts	Costar, Badhoevedorp (NL)
Glutamine	Seromed, Berlin
Nylon mesh	ZBF, Zürich (CH)
Trypsine	Seromed, Berlin

All companies are based in Germany except A = Austria, CH = Switzerland; NL = the Netherlands. Additional materials for special applications (*see*: Specific Applications).

M199 without Phenol red	Sigma, Deisenhofen
Fluoresceine	Sigma, Deisenhofen
FITC-dextrans (MW 4–40 kDa)	Sigma, Deisenhofen

Methods

Required media and solutions

Preparation medium

Medium M199 with penicillin/streptomycin (200 μg/ml) and 0.7 mM glutamine.

Plating medium

Medium M199 with penicillin/streptomycin (100 μg/ml), gentamicin (100 μg/ml), 0.7 mM glutamine and 10% ox serum.

Culture medium

Medium M199 with penicillin/streptomycin (100 μg/ml), 0.7 mM glutamine and 10% ox serum.

PBS^{++}

Phosphate buffered saline containing Ca^{2+} and Mg^{2+}.

PBS^{--}

Phosphate buffered saline without Ca^{2+} and Mg^{2+}.

Figure 1 Homogenising the brain tissue using a cutter with rolling blades.

Isolation of brain capillary endothelial cells

Brains are removed from freshly slaughtered 6-month-old pigs and placed in ice-cold 70% ethanol for some minutes. For transport to the laboratory ethanol is substituted by 4 °C cold phosphate buffered saline (PBS^{--}, i.e. without Ca$^+$ and Mg^{2+} but containing 200 U/ml of penicillin and 200 μg/ml of streptomycin).

Brains are then transferred into fresh PBS^{++}. Storage of brains for up to 4h is possible although an immediate use is recommended to avoid a loss of viability and to obtain a high yield of isolated cells. After a short flaming of the whole brains, the meninges are removed carefully and completely. Secretory areas (plexus choroideus, thalamus, etc.) are removed. The obtained tissue containing white and grey matter is minced using a sterile cutter with staggered rolling blades (Figure 1). The homogenate is collected in the preparation medium and suspended by a magnetic stirrer. The yield of one brain is diluted to a final volume of 100 ml.

Dry-powdered dispase II at 1% (w/v) (neutral, non-specific protease from *Bacillus polymyxa*) is added, followed by moderate stirring for about 3 h at 37 °C. A dextran solution (150 ml, MW 160 kDa, 18% w/v) is added per 100 ml of digested solution to obtain a final 10.8% suspension. After mixing this suspension it is centrifuged for 10 min at 6800 g and 4 °C. The supernatant is discarded to collect the vessel-containing pellet. For separation of remaining larger vessels the pellet is resuspended and passed through a 180 μm nylon mesh. The filtrate is diluted with the preparation

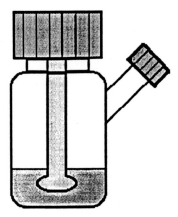

Figure 2 Flask with propeller stirrer to avoid mechanical stress.

medium to a volume of 10 ml/brain and adding 0.1% w/v collagenase/dispase II (from *Vibrio alginolyticus/Bacillus polymyxa*). The following digestion is performed at 37 °C for about 2–3 h under gentle stirring with a hanging magnetic stirrer.

The use of this magnetic stirrer (Figure 2) is a critical step in the preparation. Samples have to be taken for microscopic inspection since the activity of the commercially available enzyme mixture varies from batch to batch. Moreover, the stirring process itself is critical. A conventional magnetic stirrer is not recommended since it causes some shear stress. We used a propeller stirrer with a minimum speed just high enough to avoid the formation of a pellet.

Released cell aggregates are collected by low-spin centrifugation (140 g, 10 min., 20 °C). For further purification cells obtained from 1–2 brains are resuspended in 10 ml of preparation medium and centrifuged on a discontinuous Percoll-gradient prepared from 15 ml 1.07 g/cm^3 and 20 ml 1.03 g/cm^3. Centrifugation is carried out at 1300 g for 10 min in a swing-out bucket rotor. The cellular fraction appearing at the interface is carefully aspirated in a minimal volume avoiding shear stress. Cells are further washed by suspension in the preparation medium and a centrifugation at 140 g for 10 min at room temperature. Cells obtained here are aggregated to clusters originating from former capillaries.

Cell culture

Cell clusters from one brain are sown on 250 cm^2 culture surface in the plating medium. Collagen-coated culture surfaces allow reduction of plating density to 500 cm^2 per brain. Two days after sowing, cells are washed twice with PBS^{++} to remove debris and non-adherent blood cells and fresh culture medium is added. Further exchange of culture medium is performed every 3 days. Cells reach confluence after 7 days if uncoated surfaces are used and 5 days on collagen-coated surfaces.

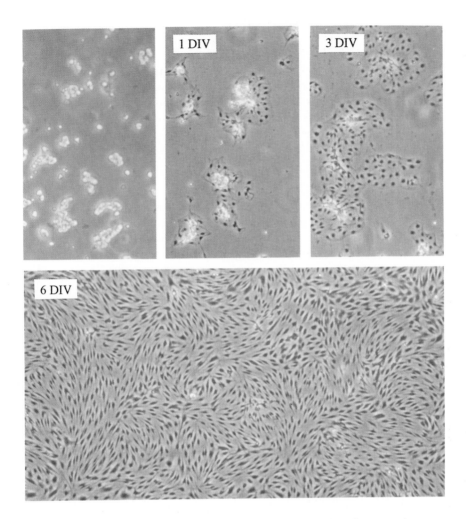

Figure 3 BCECs after different days of cultivation (DIV = days *in vitro*).

Since confluent cell layers tend to dissociate from the culture surface, cultivation time is limited to 14 days.

Subcultivation of cells

Practical applications demand for a perfect monolayer free of capillary fragments and contaminating pericytes. Thus we perform a subcultivation method. Cell clusters are sown on collagen-coated dishes. After reaching about 70% confluence the cells are treated with 0.02% trypsin at room temperature for 5–10 min. In this early step of trypsination proliferating cells descending from cell clusters are detached while cell clusters and pericytes remain on the culture surface. The process of detachment has to be continuously followed by light microscopy to determine the optimal incubation time. The volume is doubled by the addition of 0.02% EDTA and the suspension is incubated for 15 min. at 37 °C. Afterwards, trypsine activity is inhibited by ox serum. In order to obtain a single cell suspension, the cells are passed through a 64 μm nylon mesh. Finally cells are collected by centrifugation (140 g, 10 min) and sown on about 60% of the former culture surface area. Note that the subcultured cells have approximately the same proliferation rate as the primary cultured cells.

Transport study

In order to study transport, permeability and electrical resistance of primary or subcultivated cells, they are sown on translucent polycarbonate Transwell™ inserts coated with rat tail collagen. Care has to be taken while washing and replacing the medium, since the filter inserts are fragile and cells tend to detach due to shear stress.

SPECIFIC APPLICATIONS

The development of a transendothelial resistance (TEER) indicating a 'tight' monolayer is the prerequisite to study transport, permeability and their modulation. In addition to the conventional whole filter method for resistance measurements we developed two new applications to gain more specific information on TEER. The first is a device using a diminished measuring area, smaller than 1 mm², allowing to detect heterogeneities in TEER at different locations of the cell monolayer (Erben *et al.*, 1995). The second method is based on AC impedance analysis of the TEER. Advantages of this application are the possibility of long-term studies under normal cultivation conditions, performing highly resolved time scans and the opportunity to measure TEER on cell monolayers grown on impermeable supports (gold electrodes). For detailed information see references (Wegener *et al.*, 1996).

The size-dependent permeability is another important parameter to prove the integrity of the confluent cell monolayer. Therefore we carried out permeability studies using a non-radioactive assay.

Filters with confluent cell layers are placed in a suitable glass compartment. Fluoresceine and FITC-conjugated dextrans of different molecular weights are applied to the apical site of the Transwell™ insert. As phenol red quenches the fluorescence of fluorescein the use of phenol red free media is recommended. Samples are collected every 20 min from the basolateral compartment which is stirred in order to avoid a diffusion gradient and the fluorescence at 530 nm is measured with a microplate reader.

From these studies we recommend that endothelial cell monolayers used in transport studies have to exhibit a TEER of at least 100 $\Omega \cdot$cm² and a permeability of less than 1% clearance for 4 kDa dextran in 2 h.

PROBLEM SOURCES, QUALITY CONTROL AND CONCLUDING REMARKS

Due to the dedifferentiation of BCEC *in vitro* we exclusively used primary cell cultures or the first subculture. Isolated cells have to be characterized extensively on their common endothelial markers such as the expression of factor VIII related antigen, angiotensin-converting enzyme and the uptake of DiI-Ac-LDL. Furthermore the isolated cells should provide specific BBB characteristics, high specific activity of ALP and γ-GT, and considerable TEER values. Electrical resistances of confluent BCEC layers, obtained by the presented method, are detected in a range of 50 to 130 $\Omega \cdot cm^2$ with an average maximum of 120 $\Omega \cdot cm^2$, verified by the methods mentioned above (Erben *et al.*, 1995; Wegener *et al.*, 1995).

The main problem in the cultivation of endothelial cells derived from mircovessels is the contamination with pericytes. Growing between endothelial layer and culture surface pericytes may provoke leakages and cause a breakdown of the TEER and an increase in permeability. Pericyte-contaminated cultures should be purified by the additional subcultivation step described above.

REFERENCES

Bowman, P.D., Ennis, S.R., Rarey, K.E., Betz, A.L. and Goldstein, G.W. (1983) Brain microvessel endothelial cells in tissue culture: A model for studying the blood-brain barrier permeability. *Ann. Neurol.*, **14**, 396–402.

Mischeck, U., Meyer, J. and Galla, H.-J. (1989) Characterisation of the γ-glutamyl transpeptidase activity of cultured endothelial cells from porcine brain capillaries. *Cell. Tiss. Res.*, **256**, 221–226.

Erben, M., Decker, S., Franke, H. and Galla, H.-J. (1995) Electric resistance measurements on cerebral capillary endothelial cells: A new technique to study small surfaces. *J. Biochem. Biophys. Methods*, **30**, 227–238.

Wegener, J., Sieber, M. and Galla, H.-J. (1996) Impedance analysis of epithelial and endothelial cell monolayers cultured on gold surfaces. *J. Biochem. Biophys. Methods*, (in press).

V.1. ISOLATION AND CHARACTERISATION OF HUMAN BRAIN ENDOTHELIAL CELLS

M.S.F. CLARKE,[1] D.C. WEST,[1,3] P. DIAS,[2] S. KUMAR AND P. KUMAR

Christie Hospital, Manchester M20 9BX, UK

The availability of tissue cultured endothelial cells (EC) has immensely added to our understanding of vascularisation (i.e. angiogenesis), the blood brain barrier (BBB), thrombogenesis and the pathobiology of vascular diseases. Most of these studies have been undertaken utilizing EC from large vessels such as human umbilical vein EC. These are not ideal cells for developing an *in vitro* model as they represent only 1% of the total vascular area *in vivo* and there are profound differences between large and micro-vessel EC. Hence it is more appropriate to use microvessel EC. It is also important to examine how EC of pathological tissues compare with their normal counterparts. Here we have evaluated a number of methods for the isolation of microvessel EC. A detailed description is given of the characterisation and development of *in vitro* models to examine the effects of angiogenesis promoters and inhibitors. We have used these to develop EC antibody directed therapy in patients with brain tumours. Similar strategies can be used after stroke, myocardial infarction and radiotherapy.

INTRODUCTION

The culture of human endothelial cells (EC) from large vessels has been carried out by a number of investigators. However, it is more appropriate that microvessel derived endothelial cells (MEC), rather EC from large vessels, be used for the investigation of EC associated diseases. Isolation of MEC is generally a two-stage process involving tissue homogenization followed by preparation of the capillary vessel portion. One of the major problems associated with the isolation of human EC is the degree of purity.

Several techniques have been employed to minimise contamination which include the physical removal of contaminant cells from EC isolates. Using a step-wise Percoll gradient, Pertoft and Laurent (1982) achieved EC separation from a mixed cell isolate. This method was modified by Sbarbati (1984) to isolate MEC from enzymatically digested skin biopsies using a continuous rather than step-wise Percoll gradient. Selective outgrowth procedures have also been used in the preparation of pure MEC cultures (Carson *et al.*, 1989). Methods to purify EC which depend on selective killing, or growth inhibition of contaminant cells, have been reported. For instance toxic radioactive thymidine, which is taken up at a greater rate by the faster growing contaminant cells, is a technique referred to as 'thymidine suicide' (Schwartz, 1978). The use of growth supplements, such as heparin, which appears to promote EC growth while inhibiting smooth muscle cell proliferation (Thornton *et al.*, 1983) has proved helpful in the control of contaminating cells. Other

[1]Present address: Department of Immunology, Liverpool University, UK. [2]Pharmingen, San Diego, USA.

[3]Correspondence: D.C. West, Department of Immunology, Faculty of Medicine, University of Liverpool, P.O. Box 147, Liverpool L69 3BX, UK.

substances, like progesterone and medroxyprogesterone, are toxic to fibroblasts at concentrations of 1.4×10^{-4} to 1.4×10^{-3} M *in vitro* (Andrada *et al.*, 1985). Spermine, a polyamine, is selectively toxic to fibroblasts *in vitro* (Jensen and Therkelsen, 1982). The ability of these compounds to selectively inhibit fibroblasts in EC cultures is still under investigation.

After isolation of a primary EC it is important to establish that the isolated cells are actually EC. Morphological criteria may prove adequate for the initial screening but are certainly not definitive, especially with respect to MEC. Therefore, the cells must be identified by the presence of specific EC markers. The marker of choice in human EC identification had been the von Willebrand Factor (vWF). This antigen is expressed in human EC *in vivo* and *in vitro* (Jaffe *et al.*, 1973). Recently the use of the vWF as a marker of cultured EC has been shown to be of limited value in some situations as this antigen may be lost during culture or even be absent in some EC, (Rosen and Goodman, 1987; Wang *et al.*, 1994). Nevertheless, it is a useful EC marker (Kumar *et al.*, 1987). Selective uptake of acetylated low density lipoprotein (Dil-Ac LDL) is another reliable aid for the identification of EC. Several monoclonal antibodies (Mabs) which recognise EC specific antigens (e.g. CD31, CD34, CD105) have been used to identify human EC in culture (Kumar *et al.*, 1995).

MATERIALS AND METHODS

Tissue culture flasks were obtained from Costar, Cambridge, MA., USA, and basal media, Dulbecco's modification of Eagles medium (DMEM) and Medium 199 (M 199) were from Flow Laboratories, Irvine, Scotland and Medium CDB 131 (MCDB 131) was obtained from Gibco Ltd, Paisley, Scotland. Trypsin and glutamine were purchased from Flow Laboratories, Irvine, Scotland. Penicillin and streptomycin from Wellcome Laboratories, UK. Dimethylsulphoxide (DMSO), heparin, collagenase (Type II), fibronectin (FN) (bovine origin, lyophilised, cell toxicity tested), bovine serum albumin (BSA) (Fraction IV, cell toxicity tested), endothelial cell growth factor (ECGF) (bovine origin, 95% pure, lyophilised), spermine and progesterone were all purchased from the Sigma Chemical Company Ltd, Dorset, England. [^{3}H]-thymidine (tissue culture grade) and [^{35}S]-L-methionine (tissue culture grade) were obtained from Amersham International plc, Aylesbury, Buckinghamshire, UK.

Six different batches of foetal calf serum (FCS) from 3 suppliers (Flow Laboratories, Sigma and Gibco) were assessed for their ability to support human EC growth. This was carried out using [^{3}H]-thymidine incorporation as a measurement of cell growth. Briefly, EC were seeded at a density of 10,000 cells/well in replicates of three, into FN-coated 24-well culture plates. Cells were grown in 10% FCS.DMEM (FCS having been decomplemented by heating at 56 °C, for 30 min) containing 100 IU/ml penicillin, 100 μg/ml streptomycin, 300 μg/ml glutamine and 10 μCi/ml [^{3}H]-thymidine label and incubated for 5 days at 37 °C, in a 5% CO_2/air atmosphere. [^{3}H]-thymidine incorporation (CPM/well was assayed and an average value was obtained for each FCS batch. The FCS batch which yielded the greatest

[^{3}H]-thymidine incorporation was used throughout this study for the culture of human EC.

Fluorescently labelled Dil-Ac-LDL was obtained from Biogenesis Ltd,Portsmouth, England; fluorescein-conjugated *Ulex europeus* agglutinin I (UEA-I) and IgG fluorescein-conjugated rabbit anti-mouse antibody (RαM-FITC) were purchased from Dako; α-L-Fucose and Hoechst fluorescent dye was purchased from Sigma. All other standard laboratory reagents used were of Analar^R or higher grade.

Large vessel EC were isolated from two different sources, umbilical (HUVEC) and saphenous vein (HSVEC). The method used was a modification of the technique developed by Jaffe *et al.* (1973). Human umbilical cords, obtained within four hours of delivery or human saphenous veins, obtained within two hours of surgical resection, were transported in sterile PBS containing 1 mM CaCl$_2$ and 1 mM MgCl$_2$. Vessels were thoroughly washed with warm balanced salt solution (BSS), containing 200 IU/ml penicillin and 200 μg/ml streptomycin, and then flushed with warm BSS to remove any remaining blood. A sterile large bore needle (internal diameter 1.5 mm) was introduced into one end of the vessel and clamped in position. A PBS solution, containing 1 mg/ml collagenase (Type II, Sigma), was introduced into the lumen of the vein, and the other end of the vessel was clamped, ensuring that there was no leakage of the collagenase solution. In the case of saphenous vein any collateral vessels were tied off using sterile suture thread. The vessel was incubated in warm, sterile BSS at 37 °C for 30 min. The collagenase solution was collected and the vessel flushed through with 20 ml of 10% FCS.DMEM, along with gentle massaging of the outside of the vessel, to remove any remaining EC. Cells were collected by centrifugation at 1000 g for 10 min, and the pellet was resuspended in 10% FCS in MCDB 131. The cell suspension was plated into fibronectin-coated T25 culture flasks (Costar) [flasks coated with PBS containing 10 μg/ml fibronectin (FN) for 4 h at 37 °C] or purified using an immunoadsorption technique (see below).

Human brain endothelial cells (HBEC) were isolated from fresh brain tissue. The brain was transported in ice and was washed in Betadine^{TM} to remove any surface contaminants. This was washed off with PBS, at 37 °C, under sterile conditions. The grey matter was removed and the remaining white matter was minced into small pieces, using a pair of sterile scalpels and homogenised with a teflon homogeniser (clearance 0.025 mm, at 400 rpm) using 10 up and down strokes. The tissue homogenate was suspended in BSS at 37 °C and filtered through a sterile 150 μm nylon mesh filter to remove large vessels and fibrous material. The filtrate was filtered through a sterile 75 μm nylon mesh to trap microvessels. These were washed in DMEM containing 15% foetal calf serum, 100 IU/ml penicillin, 100 μg/ml streptomycin, 300 μg/ml glutamine and 100 μg/ml of EC growth factor (ECGF). The suspended capillary vessels were plated onto 35 mm petri dishes (Costar) and incubated for 4 days to allow the attachment of EC. Soon after, distinct cell colonies could be observed, those which exhibited EC morphology were ring-cloned and inoculated into 48-well plates containing 15% FCS, DMEM plus 100 μg/ml ECGF. These cultures were incubated for a further 3 days and expanded into 24-well plates followed by 35 mm petri dishes and finally into T25 culture flasks. These cells were further purified using an immunoadsorption technique (see below).

To identify cells they were grown on glass cover-slips in Leighton tubes for 3 days. Cells were washed three times with warm PBS and fixed with, either ice-cold acetone for 4 min, or PBS containing 0.05% glutaraldehyde for 30 min at room temperature. Acetone-fixed cells were allowed to air dry in preparation for staining with anti-vWF monoclonal antibody (Mab used at a 1/500 dilution of ascites fluid, kindly provided by Dr. A. Goodall, Royal Free Hospital, London), whereas glutaraldehyde-fixed cells were washed three times with PBS over 10 min at room temperature, after which time Mab 5.6E (CD31), (1:1000 dilution of ascites fluid produced by Drs. P. Dias and S. Kumar) was applied directly to the cells. The cells were incubated with primary antibody for 60 min at room temperature and washed three times with PBS over 10 min. This was replaced with RαM-FITC (1/40 dilution) and incubated for a further 30 min at room temperature. The cover-slips were washed three times with PBS over 10 min, mounted in Citifluor™ on a glass microscope slide and viewed with a Leitz Dialux 22 EB fluorescent microscope.

To investigate the uptake of Dil-Ac-LDL, EC were grown on cover-slips in Leighton tubes for 3 days after which time serum-free DMEM, containing 10 μg/ml Dil-Ac-LDL was introduced, incubated for 4 h and the cells were fixed in a PBS containing 0.05% glutaraldehyde, for 30 min at room temperature. The cover-slip was washed three times in PBS, over 10 min, and mounted in Citifluor™ on a glass microscope slide. Cells were viewed with a Lietz Dialux 22 EB fluorescent microscope. Positive cells stained with cytoplasmic vesicular staining.

For staining with UEA, cells were cultured on cover-slips in Leighton tubes for 3 days, washed three times with warm PBS and fixed in a PBS solution, containing 0.05% glutaraldehyde, for 30 min at room temperature. Cells were washed with PBS containing 0.0001% CaCl$_2$ and 0.0001% MgCl$_2$ (dPBS) for 10 min which was replaced with a 1/100 dilution of FITC-linked UEA-1 lectin in dPBS and incubated for a further 30 min at room temperature. The cells were washed in PBS and mounted in Citifluor and viewed with a Leitz Dialux 22 EB fluorescent microscope. EC exhibited a diffuse staining pattern. The specificity of the lectin staining was determined by the use of α-L-fucose to inhibit specific staining.

Since prostaglandin synthesis can be used as a secondary means of identifying endothelial cultures, cells were plated (100,000 cells/well) in 12-well culture plates. After a 48 incubation, 100 μl of culture medium were removed from each well and assayed for the presence of 6-keto-prostaglandin F$_1$ to determine the basal level of prostacyclin production. This was carried out by radioimmunoassay (Amersham).

In order to remove contaminant cell types from primary EC isolates several methods were tried including a novel purification immunoadsorption technique which is described here.

The monoclonal antibody, Mab 5.6E, recognises CD31 which is found in human EC but not fibroblasts or smooth muscle cells. EC from tissues or trypsinised primary cultures, were resuspended in serum-free MCDB 131 (sf MCDB131) containing 4% BSA (4%BSA.sf MCDB 131), ensuring a single cell suspension. The suspension was inoculated into a T75 culture flask (Costar) (10,000 cells/cm^2 of culture surface), which had been coated with 10 ml PBS solution containing Mab 5.6E for 4 h at 37 °C.

TABLE 1 Efficacy of various tissue culture media to support the growth of endothelial cells.

Basal Medium	Supplier	% [³H]-Thymidine Incorporation (mean ± S.D.)
DMEM	Flow	100 ± 6
M 199	Flow	154 ± 4
MCDB 131	Gibco	162 ± 13

Prior to inoculation, the flasks had been incubated with 4% BSA in sf MCDB 131 for 20 min at 37 °C to minimise non-specific binding. Cells were allowed to attach for 40 min at 37 °C and the culture surface was washed five times with serum-free MCDB 131 to remove any unattached cells. Serum-free MCDB 131 was replaced with 10 ml of 10% FCS.MCDB 131 and the cells were incubated overnight. After this time, the cells were again washed with serum-free MCDB 131, replaced with 10%FCS.MCDB 131 and cultured as normal.

RESULTS

The variability encountered between different foetal calf serum batches, with regard to their ability to support human EC in culture, was investigated employing the ³H-thymidine incorporation assay as a measure of cell proliferation. Using this technique a foetal calf serum batch was identified, which supported the growth of human endothelial cells, without the need for additional growth supplements such as ECGF. The results of 3 different basal media in their ability to support the growth of EC in culture is shown in Table 1.

Both M 199 and MCDB 131 supported enhanced levels of cell growth over that seen with DMEM. However, MCDB 131 proved the better medium with regard to passaging efficiency, allowing up to twelve passages without a significant decrease in growth rate, unlike M 199 which had a variable ability to support passaging in culture (data not shown).

In an effort to remove contaminant fibroblasts from primary cell cultures their growth media were supplemented with the polyamine, spermine, or progesterone. Spermine, at a concentration of 10 μg/ml of medium, showed no significant decrease in fibroblast contamination after 5 days, as determined by morphological criteria and fluorescent immunohistochemistry. Also, this agent appeared to decrease the rate of EC growth, since in the presence of spermine, confluence was never achieved. Progesterone, at a concentration of 1.4×10^{-3} M over a 4 h period, induced rapid cell death. Cells were observed to detach from the culture surface within 30 min. Detachment was preceded by morphological changes and the consequent loss of cell-cell adhesion. However no significant differences in the effects of this agent on EC and contaminant fibroblasts was observed, as treated and untreated cultures appeared to be identical on morphology and fluorescent immunohistochemistry.

The use of immunoadsorption proved to be most successful in obtaining pure cultures of EC. Cell attachment to the coated surface was observed within 20 min of plating and the cells had completely flattened within 40 min. The remaining loosely attached, or floating cells, were removed from the flask by washing five times with serum-free MCDB 131 and the adherent cells were cultured in 10% FCS.MCDB 131.

In order to quantify the selectivity of the binding of EC to Mab 5.6E-coated immunoadsorption plates it was necessary to devise an assay capable of measuring EC binding in the presence of fibroblasts. Human EC and adult human skin fibroblasts (AHF) were radioactively labelled by incubation in 10% FCS.MCDB 131, containing 10 μCi/ml [^{35}S]-L-methionine, for 12 h at 37 °C. Cells were trypsinised, collected by centrifugation (1000 g for 10 min), resuspended in medium and cell number/ml determined. A cell pellet containing 5000 labelled cells was digested in 1 ml of 0.1 M NaOH. An aliquot of this digest (0.4 ml) was added to 4 ml scintillation fluid and the radioactivity (CPM/5000 cells) was determined. Unlabelled human EC and fibroblasts were prepared in the same manner i.e. 5000 cells/0.5 ml serum-free MCDB 131 containing 4% BSA. 24-well plates (Costar) had previously been coated with a PBS solution containing Mab 5.6E. Cells were inoculated into either 5.6E-coated 24-well plates, FN-coated plates or untreated plates, in replicates of 3, in the following manner; labelled fibroblasts (5000 cells/well), labelled EC (5000 cells/well), labelled EC (5000 cells) plus unlabelled fibroblasts (5000 cells) (total 10,000 cells/well), unlabelled EC (5000 cells) and labelled fibroblasts (5000 cells) i.e. a total of 10,000 cells/well.

The plates were incubated for 40 min, after which each well was washed five times with serum-free media to remove any non-adherent cells. Each well was washed three times with warm PBS and the cells solubilized *in situ* with 1 ml of 0.1 M NaOH for 30 min at 37 °C. Each digest at 0.4 ml was added to 4 ml of scintillation fluid and the radioactivity per well determined.

The results from this assay are presented in Table 2 and show that EC are selectively immunoadsorbed onto 5.6E-coated wells whereas fibroblasts are not. In the presence of fibroblasts, the percentage of EC binding, although lower than in the case of EC alone, is again shown to be significantly increased over the control values. The control values, i.e. labelled fibroblasts in the presence of unlabelled EC and the medium blank containing 10 μCi/ml [^{35}S]-L-methionine, are seen not to be significantly different from that of labelled fibroblasts alone. This indicates that Mab 5.6E immunoadsorption of EC is highly specific, even in the presence of non-EC types.

Fibronectin, although enhanced the plating efficiency of EC, did not show any EC specificity with regard to cell attachment. This is illustrated by the parallel increase in fibroblast and EC percentage binding to FN-coated plates. Untreated plastic plates were also assayed in this manner and showed an increase in fibroblast rather than EC binding.

The identification of a FCS batch which had the ability to support human EC growth in culture, without the need for additional growth supplements, allowed the comparison of different basal media preparations for their ability to support human EC growth using serum alone, as determined by the [^{3}H]-thymidine incorporation

TABLE 2 Binding of human endothelial and fibroblast cells to untreated, FN-coated and Mab 5.6E-coated culture surfaces.

Treatment	Cell type				
	EC*	AHF*	EC* & AHFᶜ	ECᶜ & AHF*	Control
Plastic	5%	52%	4%	37%	≤ 5%
FN	62%	73%	44%	58%	≤ 5%
5.6E (CD31)	83%	7%	52%	6%	≤ 5%

EC* – labelled endothelial cells, AHF* – labelled fibroblasts, ECᶜ – unlabelled endothelial cells, AHFᶜ – unlabelled fibroblasts, Control – media blank.

assay. Why MCDB 131 (Knedler and Ham, 1987) has proved so successful in the culture of human EC using relatively low serum concentrations alone is unknown. The increased Mg^{2+} ion availability in this medium may play a role as this has been implicated in cell activation (Rubin *et al.*, 1978). It is worth noting that freshly prepared MCDB 131 gave better results than older medium which has been stored. When compared to MCDB 131, M 199 was less efficient in supporting passaging efficiency of older cultures.

Once the initial problems with regard to establishing human EC in culture had been overcome, further problems of the presence of non-EC types, such as fibroblasts and smooth muscle cells, in the primary endothelial isolates had to be overcome. The use of spermine (Jensen and Therkelsen, 1982) and progesterone (Andrada *et al.*, 1985) as selectively toxic agents against contaminant fibroblasts in primary epithelial cell cultures suggested that they may be used for a similar purpose in primary human EC cultures. Unfortunately both agents displayed toxicity to EC which appeared to be as great, if not greater, than it was for fibroblast.

The Mab-mediated removal of leukaemic cells during bone-marrow purging (Reading *et al.*, 1985), suggested that this approach to separating a mixed cell population could also be applied to the purification of EC. Indeed this proved to be the case.

In an attempt to quantify human EC binding to Mab 5.6E coated culture surfaces, a simple radiometric assay was developed. This employed [^{35}S]-L-methionine labelled human EC and adult human fibroblasts to study cell binding to Mab 5.6E coated surfaces. Mixtures of these two cell types were also used in an effort to mimic the mixed cell populations found in primary cell isolates. Cell binding to untreated plastic and FN-coated culture surfaces was also investigated in a similar manner. In the case of plastic culture surfaces, EC binding was very low compared to that observed with AHF. FN-coating resulted in a general increase in both endothelial and AHF cell binding, a result consistent with the fact that it provides a partial extracellular matrix, so enhancing cell attachment. In the case of Mab 5.6E-coated culture surfaces, AHF cell binding, alone or in the presence of unlabelled EC, was not significantly different from those of the cell-free medium blank. On the

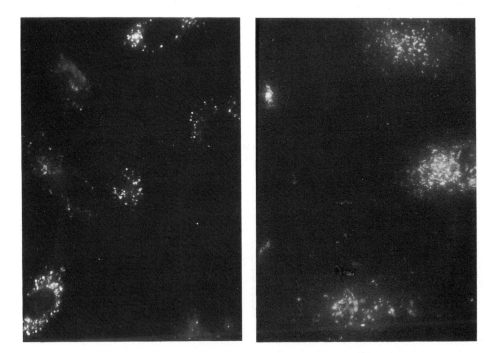

Figure 1 Characterisation of purified human umbilical vein endothelial cells using two classical markers (for details see text and Kumar *et al.*,1987). (left) Dil-AC LDL and (right) von Willebrand factor positive cells.

other hand, EC binding to Mab coated surfaces was greatly enhanced, even in the presence of unlabelled AHF. The decrease in EC binding observed in the presence of a contaminant cell type, over the levels seen as a pure cell culture, is probably due to physical impediment of attachment by unlabelled cells. These results illustrate the highly specific nature of this novel immunoadsorption technique for the purification of human cultures. One further point which can be made with regard to the high purity gained using this technique is that, although to claim 100% purity is tenuous, cultures treated in this manner have been cultured up to passage 12 without the appearance of non-endothelial cell types based either on morphological, immunohistochemical or biochemical criteria (Table 3 and Figure 1). This method of purification has been used to establish pure cultures of HUVEC, HSVEC and HBEC and a number of other human types, including iliac and femoral artery, subcutaneous adipose microvascular and bone marrow EC.

The actual mechanism involved using Mab 5.6E immunoadsorption may be a simple selective adhesion of EC to a specific extracellular matrix. However, it has been suggested that Mabs can be growth inhibitory to EC (A. Poot, personal communication). This is not the case with Mab 5.6E, as immunoadsorbed cells can

TABLE 3 Staining patterns of endothelial cells isolated by Mab 5.6E immunoadsorption.

Cell Type	Marker				
	vWF	DiL-Ac LDL Uptake	5.6E (CD31)	UEA-1	Prostacyclin (ng/10,000 cells)
HUVEC	+	+	+	+	5.4
HSVEC	+	+	+	+	3.6
HBEC	+	+	+	+	8.3
AHF	–	–	–	–	0

(–) negative; (+) positive; (±) weak/inconclusive staining; (ND) not done, HUVEC: in human umbilical vein endothelial cells; HSVEC: human saphenous vein endothelial cells; HBEC: human brain endothelial cell; AHF: human fibroblasts.

be grown to confluence on the coated culture surfaces and then passaged back onto FN-coated surfaces with no apparent detrimental effects. Conversely, EC grown on Mab 5.6E coated surface did not show any increase in growth rate compared to cells grown on fibronectin. Also, the presence of free Mab 5.6E did not affect EC growth rate as determined by [^{3}H]-thymidine incorporation. Initial experiments with regard to the effects of Mab 5.6E on AHF cultures have yielded some surprising results. AHF can attach to Mab 5.6E coated surfaces if left for extended periods of time (approximately 8 h). This attachment appears to be only transient and the cells lift off again and remain in suspension.

Immunoadsorption has several advantages over the existing methods for human EC purification, such as cloning/farming (Carson *et al.*, 1989), in that it yields a more representative EC cell population than an EC culture derived from a single cell. By pooling purified human EC of the same type from a number of isolations, inter-individual EC differences have to some extent been overcome.

SPECIFIC APPLICATIONS

We have isolated endothelial cells from normal human brain and brain tumour microvessels. These were used to develop an *in vitro* model for EC directed therapy in patients with brain tumours. The technique applied to brain tumour therapy i.e. anti EC antibodies can be applied to other angiogenic diseases such as stroke, myocardial infarction, rheumatoid arthritis, diabetic retinopathy, etc.

CONCLUDING REMARKS

The availability of EC specific antibodies, growth factors, lectins and flow cytometry has revolutionized techniques to obtain a large number of endothelial cells from microvessel of various tissues, including the brain. Here we have alluded to their uses as *in vitro* model to study blood brain barrier.

REFERENCES

Andrada, E.C., Hoschoian, J.C., Anton, E. and Lanari, A. (1985) Growth inhibition of fibroblasts by progesterone and medroxprogesterone *in vitro. Int. Archs Allergy Appl. Immuno.*, **76**, 97.

Carson, M.P. Saenz-de-Tejada, I., Goldstein, I. and Haudenschild, C. (1989) Culture of human corpus cavernosum endothelium. *In Vitro*, **23**, 248.

Jaffe, E.A., Nachman, R.L. and Becker, C.G. (1973) Culture of human endothelial cells derived from human unbilical veins; identification by morphologic and immunologic criteria. *J. Clin. Invest.*, **52**, 2745.

Jensen, P.K.A. and Therkelsen, A.J. (1982) Selective inhibition of fibroblasts by spermine in primary cultures of normal human skin epithelial cells. *In Vitro*, **18**, 867.

Knedler, A. and Ham, R.G. (1987) Optimized medium for clonal growth of human microvascular endothelial cells with minimal serum *in vitro. Cellular and Development Biol.*, **23**, 481.

Kumar, S., West, D.C. and Ager, A. (1987) Heterogeneity in endothelial cells from large vessels and microvessels. *Differentiation*, **36**, 57.

Kumar, S., Wang, J.M. and Bernabeu, C. CD105 and Angiogenesis. *J. Path.*, (in press).

Pertoft, H. and Laurent, T.C. (1982) Isopycnic separation of cells by centrifugation in Percoll gradients. In: *Cell Function and Differentation, Part A:*, **95**, AR Liss, NY.

Reading, C.L. and Takane, Y. (1985) Monoclonal antibody applications in bone marrow transplantation. *Biochem. Biophys. Acta*, **865**, 141.

Rosen, J.D. and Goodman, A.L. (1987) Heterogeneity of rabbit aortic endothelial cells in primary culture.*Proc. Soc. Exp. Biol. Med.*, **184**, 495.

Rubin, A.H., Terasaki, M. and Sazui, H. (1978) Magnesium reverses inhibitory effects of calcium deprivation on coordinate response of 3T3 cells to serum. *Proc. Natl. Acad. Sci. USA*, **75**, 4379.

Sbarbati, R. (1985) Separation of human endothelial cells from fibroblasts by centrifugation in Percoll gradients. *Biosci. Rep.*, **5**, 469.

Schwartz, S.M. (1978) Selection and characterisation of bovine aortic endothelial cells. *In Vitro*, **14**, 966.

Thornton, S.C., Mueller, S.N. and Levine, E.M. (1983) Human endothelial cells; use of heparin in cloning and long term serial cultivation. *Science*, **222**, 623.

Wang, J.M., Kumar, S., Pye, D., Haboubi, N. and Al-Nakib, L. (1994) Breast carcinoma: Comparative study of tumor vasculature using two endothelial cell markers. *J. Natl. Cancer Inst.*, **86**, 386.

V.2. METHODS OF ISOLATION AND CULTURE OF HUMAN BRAIN MICROVESSEL ENDOTHELIUM

M. VASTAG AND Z. NAGY

*National Institute of Psychiatry and Neurology, National Stroke Centre,
Huvosvolgyi ut 116, Budapest, Hungary H-1021*

The technology of culturing human microvessel endothelial cells provides a new perspective in the understanding of different pathophysiological phenomena. In our method we used the enzymatic digestion of human brain cortical tissue with collagenase, and it was followed by two-step gradient centrifugation of the digested tissue sample, first in dextran and then in percoll gradient. After seeding and culturing of the separated capillary segments and individual cells, we can obtain well growing endothelial cell colonies. The endothelial cell culture can be enriched by mechanical cloning up to 95-98% and the cells can be passaged 2-3 times without losing the endothelial phenotype detected in the primary culture.

INTRODUCTION

The research on brain endothelium has been greatly enhanced by the introduction of endothelial cell culture technology. Endothelial cells are the interface between blood and tissue compartments. In the brain, microvessel endothelial cells join together with belt-like continuous tight junctions, providing ultrastructural microenvironment of the brain parenchyma (Nagy *et al.*, 1995). This cell layer is a highly selective barrier between blood and brain tissue. Normal endothelial cell layer provides a thromboresistant surface that prevents platelet or leukocyte adherence and activation of the intrinsic or extrinsic coagulation system (Curwen *et al.*, 1980; Loskutoff *et al.*, 1986). Endothelial cells are active in different physiological and pathophysiological conditions (Nagy, 1990). Their metabolic activity is highly versatile: they synthesize, among others, prostacyclin, endothelin (Rubanyi and Van Houtte, 1989), nitrous oxide (NO) (Durieu-Trautmann *et al.*, 1993), tissue plasminogen activator (Hinsberg *et al.*, 1990), plasminogen activator inhibitor (Hanss and Collen, 1987), soluble forms of adhesins and selectins (Pigott *et al.*, 1992), von Willebrand factor (Wu *et al.*, 1987). They also contain receptors for a variety of vasoactive agents. On the surface of the endothelial cells many cell surface molecules such as adhesins, selectins can be found (Stad and Buurman, 1994). The up- and down-regulation of these molecules can be of clinical importance.

To study the function of the endothelia, different models have been developed: bolus perfusion technique (Nagy *et al.*, 1983), isolation of microvessels (Joó, 1992; Joó and Karnushina, 1973) and culturing endothelial cells from animal brains (Bowman *et al.*, 1981, 1983). Culturing endothelial cells from human brain tissue is a promising new technology to study physiological and pathophysiological phenomena in the human. Human brain microvessel endothelial cell culture techniques have been developed recently (Spatz *et al.*, 1980). Modifications of enzymatic digestion with gradient centrifugation and simple mechanical separation have been used in the different laboratories although a standardised procedure is not yet available.

In this chapter we described a method of isolation and culture of human brain microvessel endothelium, which is a modification of previously described procedures (Bowman *et al.*, 1981, 1983; Tao-Cheng *et al.*, 1987, 1990; McCarroll, pers. comm.). This method is a reproducible and highly uniform, standard procedure to prepare endothelial cell culture from human (foetal or adult) brain tissue.

To use tissues of human origin, our laboratory has been granted a permission by the Ethical Committee of Health and Research Council, Hungary.

MATERIALS AND METHODS

Materials

FBS (foetal bovine serum), DMEM (Dulbecco's Modified Eagle medium), nutrient mixture Ham's F-12, penicillin/streptomycin solution, fungizone, collagenase (type I) were purchased from Life Technologies (GIBCO BRL), Austria, and gentamycin from Chinoin, Hungary. Endothelial mitogen (source: bovine hypothalamus) and DiI-Ac-LDL were obtained from Biomedical Technologies Inc., USA. Salsol A (physiological salt solution) and DPBS (Dulbecco's phosphate-buffered saline without Ca^{2+} and Mg^{2+}) have been prepared at the Institute's laboratory. Percoll and density marker beads were obtained from Pharmacia, Sweden. Dextran, anti-human von Willebrand factor and OVA (ovalbumin) were purchased from Sigma, Germany. Tissue culture flasks are available from Greiner GmbH, Germany.

Methods

Brain tissues were removed under sterile conditions, 3–4 h after death, and placed in cold physiological salt solution, supplemented with antibiotics (penicillin/streptomycin at 2000 units/ml and 2 mg/ml, respectively), gentamycin (0.2 mg/ml) and fungizone (2.5 μg/ml). After washing the tissue samples three times in fresh salt solution, the arachnoid membrane and surface vessels were removed, and the grey matter minced into small pieces (1 mm^3 in size) and homogenized. The homogenate was washed in DMEM (1:4 homogenate/DMEM) and centrifuged (10 min, 90 g, 20 °C). The supernatant containing myelin debris was then removed. The pellet was suspended in a 2 ml solution of 0,5% collagenase (1:1 DPBS/DMEM) per gram of tissue and incubated in a rotary shaker (30 min, 37 °C). More DMEM (1:1 suspension/DMEM) was added to the incubated tissue samples which were then homogenised (3-4 strokes) using a tissue grinder. Collagenase has subsequently washed out by three times centrifugation in DPBS (10 min, 90 g, room temperature). The final pellet was suspended in 15% dextran in DPBS (30 ml per gram of tissue) and centrifuged (30 min, 4000 g, 4 °C). Dextran gradient centrifugation removes most of the neural tissue elements located on the top layer of the tube. After centrifugation in dextran, the pellet consists mostly of capillaries and small amount of individual cells of different origins.

The resultant pellet was suspended in 35% percoll (5 ml 20× DPBS/100 ml solution), 25 ml per gram of tissue sample, and centrifuged (30 min, 20,000 g, 10 °C). Percoll gradient separates the red blood cells and neural cells from the capillaries and individual endothelial cells. Density markers (1.018, 1.033, 1.049, 1.062 and 1.087 gram/ml) can be used as external standards in a preparation parallel to that containing the sample. A standard gradient (1.018 g/ml) ran in parallel with capillaries and the individual endothelial cells. This layer was collected and washed three times by spinning (10 min, 160 g, room temperature). The final pellet was resuspended in 1 ml DPBS, and the volume brought to 8 ml with culture medium (20% FBS, 5 mg/ml endothelial mitogen, 1:1 DMEM:Ham's F-12 and penicillin/streptomycin at 100 units/ml and 100 μg/ml, respectively). The suspension was seeded in plastic flasks. The cultures have remained in a CO_2 incubator (5% CO_2, 37 °C). After 48 h the culture medium was changed (10% FBS, 5 mg/ml endothelial mitogen, 1:1 DMEM:Ham's F-12 and penicillin/streptomycin at 100 units/ml and 100 μg/ml, respectively). The cells spreading from the capillaries can be observed as early as on the 2nd day. The medium should be changed regularly (e.g. 2-3 days). The primary culture of endothelial cells can be enriched by mechanical cloning. The cells can be stored frozen.

CHARACTERIZATION OF ENDOTHELIAL CELLS

The cells were characterised by visualisation of von Willebrand factor and by the uptake of Dil-Ac-LDL.

Von Willebrand Factor

The cells were cultured on glass slides for 2-3 days. After confluency, the cells were washed in Tris-HCl (0.02 M) and fixed in cold 1:1 ethanol:aceton for 5 min. Non-specific binding sites were blocked by an incubation in 0.1% HS (heat inactivated), 5% OVA, 0.5 M NaCl in Tris-HCl, for 15 min. The cells were incubated with anti-human von Willebrand factor (developed in rabbit) diluted in 0.1% HS, TNO buffer (0.02 M Tris-HCl, 0.5 M NaCl, 5% OVA), for 30 min at room temperature. The cells were washed three times in TNO, and incubated with FITC-labelled antirabbit IgG diluted in TNO. Incubation time is 60 min, at room temperature and in darkness. After an extensive washing in Tris-HCl and finally in distillated water, the slides were coverslipped and examined under fluorescence microscopy.

Dil-Ac-LDL

The cells were seeded on glass slides and cultured for 2-3 days in the standard culture medium. 10 (μg/ml Dil-Ac-LDL has been added into the standard medium and the cells incubated in this medium for 4 h, at 37 °C in an incubator. After removing the medium the cells were washed several times with probe-free solution and visualised by fluorescence microscopy using standard rhodamin excitation/emission filter (EX 546).

RESULTS AND CONCLUDING REMARKS

The method we described is based on enzymatic digestion of the extracellular matrix with collagenase and on the separation and further purification of digested capillaries and single cells on continuous dextran and percoll gradient. The capillaries and individual cells can attach and grow on plastic. Collagen, fibronectin or gelatin do not influence significantly the cell growth. The success of the procedure depends on the age of the donor subject and the time elapsed from the death and the initiation of the procedure. The method detailed above does not result in 100% clean endothelial cell culture. Contamination is about 5-15 %. However, in primary culture, the groups of endothelial cells can be differentiated, based on their morphology, from the colonies contaminated with pericytes. Thus, an early mechanical cloning of the endothelial cells can enrich the culture up to 95-98%. Endothelial cells can be further characterised by the uptake of Dil-Ac-LDL. These cells also result in positive staining with anti-human von Willebrand factor.

REFERENCES

Bowman, P.D., Betz, A.L., Ar, D., Wolinsky, J.S., Penney, J.B., Shivers, R.R. and Goldstein, G.W. (1981) Primary culture of capillary endothelium from rat brain. *In vitro*, **17**, 4.
Bowman, P.D. *et al.* (1983) Brain microvessel endothelial cells in tissue culture, model for study of blood-brain barrier permeability. *Annual Neurology*, **14**, 396.
Curwen, K.D., Gimbrone, M.A. and Handin, R.I. (1980) *In vitro* studies of thromboresistance. *Laboratory Investigation*, **42**, 366–373.
Durieu-Trautmann, O., Födörici, C., Crôminon, C., Foignant-Chaverot, N., Roux, F., Claire, M., Strosberg, A.D. and Couraud P.O. (1993) Nitric oxide and endothelin secretion by brain microvessel endothelial cells: regulation by cyclic nucleotides. *Journal of Cellular Physiology*, **155**, 104–111.
Hanss, M. and Collen, D. (1987) Secretion of tissue-type plasminogen activator and plasminogen activator inhibitor by cultured human endothelial cells: Modulation by thrombin, endotoxin, and histamin. *J. Lab. Clin. Med.*, **109**(1), 97–104.
Hinsberg, V.W.M., Berg, E.A., Fiers, W. and Dooijewaard, G. (1990) Tumour necrosis factor induces the production of urokinase-type plasminogen activator by human endothelial cells. *Blood*, **75**, 1991–1998.
Joó, F. and Karnushina, I. (1973) A procedure for the isolation of capillaries from rat brain. *Cytobios*, **8**, 41–48.
Joó, F. (1992) The cerebral microvessels in culture, an update. *Journal of Neurochemistry*, **58**(1), 1–17.
Loskutoff, D.J., Ny, T., Sawdey, M. and Lawrence, D. (1986) Fibrinolytic system of cultured endothelial cells: Regulation by plasminogen activator inhibitor. *Journal of Cellular Biochemistry*, **32**, 273–280.
Nagy, Z., Kolev, K., Csonka, E., Pek, M. and Machovich, R. (1995) Contraction of human brain endothelial cells induced by thrombogenic and fibrinolytic factors. An *in vitro* cell culture model. *Stroke*, **26**(2), 265–270.
Nagy, Z. (1990) Blood-brain barrier and the cerebral endothelium. pp 11-31 Section I Pathophysiology of the blood-brain barrier. Edited by Johansson, B.B., Ferstrüm Foundation Series, 14, Elsevier.
Nagy, Z., Peters, H. and Hüttner, I. (1983) Charge-related alterations of the cerebral endothelium. *Laboratory Investigation*, **49**(6), 662–671.

Pigott, R., Dillon, L.P., Hemingway, I.H. and Gearing, A.J.H. (1992) Soluble forms of E-selectin, ICAM-1 and VCAM-1 are present in the supernatant of cytokine activated cultured endothelial cells. *Biochemical and Biophysical Research Communications*, **187**(2), 584–589.

Rubanyi, G.M. and Vanhoutte, P.M. (1989) Endothelium-derived contracting factors. First International Symposium on Endothelium-Derived Vasoactive Factors, Philadelphia, Pa., May 1-3, 1989.

Spatz, M., Bembry, J., Dodson, R.F., Hervonen, H. and Murray.M. (1980) Endothelial cell cultures derived from isolated cerebral microvessels. *Brain Research*, **191**, 577–582.

Stad, R.K. and Buurman, W.A. (1994) Current views on structure and function of endothelial adhesion molecules. *Cell Adhesion and Communication*, **2**, 261–268.

Tao-Cheng, J.H., Nagy, Z. and Brightman, M.W. (1987) Tight junctions of brain endothelium *in vitro* are enhanced by astroglia. *The Journal Neuroscience*, **7**(10), 3293–3299.

Tao-Cheng, J.H., Nagy, Z. and Brightman, M.W. (1990) Astrocytic orthogonal arrays of intramenbranous particle assemblies are modulated by brain endothelial cells *in vitro*. *Journal of Neurocytology*, **19**, 143–153.

Wu, Y.Q., Drouet, L., Carrier, J.L., Rotschield, C., Berard, M., Rouault, C., Caen, P.J. and Meyer, D. (1987) Differential distribution of von Willebrand factor in endothelial cells. *Arteriosclerosis*, **7**, 47–53.

PART B
IN VIVO: MICRODIALYSIS

I. INTRODUCTION

A.G. de BOER[1] and W. SUTANTO[1,2]

Divisions of Pharmacology[1] and Medical Pharmacology[2], Leiden/Amsterdam Center for Drug Research (LACDR), University of Leiden, P.O. Box 9503, 2300 RA Leiden, The Netherlands

Microdialysis is a relatively new *in vivo* procedure to measure quantitatively blood-brain barrier (BBB) functionality and BBB transport of drugs to the brain. The advantage of this method is that it allows the measurement of local drug concentrations in the extracellular fluid of the brain. The technique has advanced considerably in recent years, and requires sensitive assay procedures since drug concentrations in the dialysate are relatively low and reflect the unbound drug concentrations. Electrochemical or high sensitivity fluorescense procedures are needed for the quantification of these concentrations, although clean-up procedures are not needed in most cases.

Several aspects of this technique are still subject to further improvement, e.g. recovery, tissue damage/tissue recovery, time of the experiment to be performed, anaesthesia, stress, etc. In the measurement of drug transport through the BBB, it appears that all these aspects have to be estimated separately for each drug.

In the second part of this book microdialysis procedures in various animals are described. These procedures have been optimised, validated and are routinely used in various laboratories involved in a European Community-funded Biomedical Concerted Action programme entitled *Drug Transport across the Blood-Brain Barrier: New Experimental Strategies*.

II.1. MICRODIALYSIS AS A TOOL IN PHARMACOKINETIC AND PHARMACODYNAMIC STUDIES

M.R. BOUW, J.S. SIDHU AND M. HAMMARLUND-UDENAES

Department of Pharmacy, Division of Biopharmaceutics and Pharmacokinetics, Uppsala University, Uppsala, Sweden

Microdialysis (MD) is a powerful method in pharmacokinetic (PK) and pharmacodynamic (PD) research. The technique permits the unbound concentrations in brain, as well as in muscle and other tissues to be followed over time. A MD probe can also be placed in the bloodstream to monitor blood concentrations without the need for blood sampling. PD measurements of, for example, transmitters, can be made with MD in parallel to drug concentration measurements at the site of action, or other PD measurements can be made and compared with local drug concentrations. MD can also be used for protein binding studies both *in vitro* and *in vivo*.

From a PK/PD perspective, the central issues in MD include suitable, practical methodologies for estimating the relative recovery of the drug of interest and time aspects of equilibration between probe perfusate and the external medium. Two methods for calculating relative recovery *in vitro* and *in vivo* are discussed in comparison to the classical *in vitro* way of determining relative recovery. These methods are retrodialysis (RD) and the method of no-net flux (NNF, or 'difference' method).

The ability of MD to characterise changes in *in vivo* drug concentrations in tissues as well as in blood is also dependent on analytical sensitivity, which determines the frequency of sampling. Any presence of delay in the equilibration between probe and surrounding tissue and adhesion to the components of the MD system should be determined. Accordingly, assessment of the time aspects of dialysate and external medium equilibration is crucial for a proper evaluation in MD for PK/PD experiments.

The brain distribution of morphine and its metabolites is presently being studied with the use of venous (CMA/20) and striatal probes (CMA/12) in relation to the antinociceptive effect. Artemisinin distribution in muscle (CMA/20) and brain (CMA/12) is also studied in relation to its venous concentrations (CMA/20). Morphine and artemisinin are used in the present chapter to give examples of how PK/PD experiments with MD can be performed. These drugs, with different physicochemical properties, present two opposing scenarios in conducting MD and highlight several problems and limitations that may be encountered.

Abbreviations

AUC   Area under the blood, muscle or brain concentration-time curve
C_d Drug concentration in dialysate
C_{in} Inlet concentration of drug in the Ringer or HEPES-Ringer solution entering the probe
C_m Ringer or HEPES-Ringer solution spiked with a known drug concentration
C_n Last measured concentration in plasma, muscle or brain
C_u Calculated unbound concentrations of the drug in blood, muscle and brain
f_u Fraction unbound in plasma
i.v.   Intravenous administration
λ_n The terminal rate constant

LPR The method of low perfusion rate
MD The microdialysis method
NNF The method of no-net flux ('difference' method)
RD The method of retrodialysis ('reference' method)
$REC_{in\ vitro}$ Relative recovery of a drug *in vitro*
$REC_{in\ vivo}$ Relative recovery of a drug *in vivo*
$RL_{in\ vitro}$ Relative loss of a drug *in vitro*
$t_{1/2}$ The terminal half-life of a drug in blood, muscle and brain
PD Pharmacodynamics
PK Pharmacokinetics

INTRODUCTION

During the last decade the technique of *in vivo* MD has been recognised as a promising tool in the area of neuropharmacological research. It has been commonly employed to monitor neurochemicals in the extracellular fluid (ECF) of discrete brain regions (Ungerstedt, 1984; Hurd *et al.*, 1988; Benveniste and Huttemeier, 1990). Recently the technique has been used in pharmacological and in pharmacokinetic (PK) studies to determine the unbound drug concentrations in brain (Wong *et al.*, 1993; Malhotra *et al.*, 1994), liver (Scott *et al.*, 1990), muscle (Deguchi *et al.*, 1992; Sarre *et al.*, 1995; van Amsterdam *et al.*, 1995), subcutaneous tissue (Lönnroth *et al.*, 1991) and blood (Telting-Diaz *et al.*, 1992; Alonso *et al.*, 1995). In addition, MD is a potentially valuable tool for *in vitro* or *in vivo* protein binding studies (Dubey *et al.*, 1989; Herrera *et al.*, 1990; Ekblom *et al.*, 1992; Le Qellec *et al.*, 1994).

The strength of the MD is its ability to continuously monitor unbound drug levels in the extracellular fluid in the same animal without the removal or introduction of fluid, which may otherwise perturb homeostasis. In addition, MD creates the opportunity to reliably assess drug distribution to a target organ. The measurement of unbound concentrations in the extracellular fluid will supply us with unique PK and PD information (Lönnroth *et al.*, 1991). However, the measurement of the true unbound concentration requires methods for calculating *in vivo* drug recovery ($REC_{in\ vivo}$).

The question of how to determine $REC_{in\ vivo}$ in MD studies has long been a subject of debate. Currently, the method of retrodialysis (RD) (Wang *et al.*, 1993; Wong *et al.*, 1993), performed prior to systemic drug administration is preferred as the most practical means of estimating probe recovery. RD is based on relative degree of substance loss from the probe perfusate to the external medium. The validity of this method has been evaluated in comparison with other recovery methods such as the point of no-net flux (NNF) and low perfusion rate (LPR) (Menacherry *et al.*, 1992; Sjöberg *et al.*, 1992; Olson and Justice, 1993; Wang *et al.*, 1993).

Our application of MD centres on PK and PD aspects of drug research. One focus of interest is to pharmacokinetically describe the *profile* of the drug in the tissue of interest in relation to its plasma profile. Another is to compare the level of unbound drug in brain with the unbound concentrations in blood. The regional PKs of morphine and metabolites are being studied in parallel with their antinociceptive effect to further understand how much the delay in PD effects within the CNS vs. blood concentrations is a result of drug transport (PK) or PD events. Artemisinin, an antimalarial agent, is studied regarding its distribution to muscle and brain in relation to the plasma concentrations.

This chapter contains a detailed description of the requirements for and performance of MD as a tool in PK/PD research, with morphine and artemisinin as examples.

MATERIALS AND METHODS FOR SYSTEM ESTABLISHMENT

Microdialysis Apparatus

The MD equipment employed in our laboratory is supplied by CMA Microdialysis AB (Stockholm, Sweden). The basic components of the system include:

- an infusion pump (CMA/100)
- 1 ml and 2.5 ml syringes (Emirex®)
- micro fraction collector (CMA/140)
- a CMA/120 system for freely moving animals, and
- PE-50 probe inlet and FEP (Teflon, i.d. 0.12 mm, CMA) outlet tubing.

Microdialysis Probes

Two types of probes are employed in our studies: CMA/12 (3 mm membrane length, 400 μm i.d., 500 μm o.d., 3 μl internal volume) for brain microdialysis and the flexible probe CMA/20 (10 mm membrane length, 500 μm i.d., 670 μm o.d., 4 μl internal volume) for blood and muscle MD. Both probes consist of a polycarbonate membrane with a molecular weight cut-off of 20,000 Daltons.

Perfusion Solutions

Buffer solutions for the striatal (Ringer) and blood/muscle (HEPES-Ringer) probes are prepared in advance, filtered through a 0.45 μm PC membrane (Millipore, France) and 20 ml portions are stored at $-20\,°C$. On the experimental day, the required amount of the respective perfusion solutions are thawed and degassed by ultrasonication for 10 min prior to probe perfusion.

The composition of the two perfusion solutions are as follows:

Ringer: 145 mM NaCl, 0.6 mM KCl, 1.0 mM $MgCl_2 \cdot 6H_2O$, 1.2 mM $CaCl_2 \cdot 2H_2O$, and 0.2 mM ascorbic acid in 2 mM phosphate buffer (pH 7.4) (Moghaddam *et al.*, 1989).

HEPES-Ringer: 147 mM NaCl, 4 mM KCl, and 2.3 mM $CaCl_2 \cdot 2H_2O$ (Ringer Solution, Pharmacia AB, Uppsala, Sweden), with 0.06% N-(2-hydroxyethyl)piperazine-N'-2-ethanesulphonic acid (HEPES) and pH adjusted to 7.4 with 0.2 M NaOH.

Probe Preparation and Storage

New CMA probes are prepared according to the manufacturer's instructions. Briefly, the probes (soaked in 70% ethanol) are perfused with 70% ethanol for 5 to 10 min at a flow rate of 15 μl/min. Probes are then perfused with the respective perfusion solutions for 10 min and the probe membrane checked for the presence of air bubbles before use. During both *in vitro* and *in vivo* experiments, the probes are perfused with a flow rate of 2 μl/min, unless otherwise stated. Following a MD probe experiment, probes are perfused with distilled water to prevent crystallisation of buffer salts and stored at room temperature in distilled water. Probes are generally used 3 to 4 times and checked for recovery before each use.

Chemicals

Morphine HCl (10 mg/ml) is purchased from Pharmacia AB (Stockholm, Sweden). Artemisinin is obtained from the State Pharmaceutical Administration (People's Republic of China). Hypnorm® is obtained from Janssen Pharmaceutics, Beerse, Belgium. Chloral hydrate (36 mg/ml) is obtained from the Academic Hospital Pharmacy (Uppsala, Sweden). Low molecular weight heparin is purchased from Sigma Chemicals (St Louis, USA). All other chemicals used in the perfusion solutions and buffers are of analytical grade and are purchased from Merck. Solvents are of HPLC grade.

Morphine has a molecular weight of 285. The pK_a is 7.93. Water solubility is 57 mg/ml. Artemisinin is a neutral compound with a pK_a of 2.82. Its water solubility is low, 0.084 mg/ml and the log $P_{oct/water}$ is >4.

Animals

Male specific pathogen-free rats of Sprague Dawley descent (Møllegård, Denmark) weighing 280 to 330 g are used in all studies. The animals are acclimatised in groups in plastic cages for 4 days in the Animal Care Unit of the Biomedical Centre (Uppsala) at a temperature of 21 °C, a relative humidity of 60% and with a 12 h day-night cycle. Laboratory chow (Ewos-R3, ALAB, Sollentuna, Sweden) and tap water are available *ad libitum*. Ethical approval is obtained from the Animal Ethics Committee of Uppsala University (C313/92).

Probe Characterisation *In Vitro*

The equilibrative nature of drug transport into the probe system requires early investigation in order to evaluate a compound's candidacy for MD. Two aspects of equilibration is determined, the rate and degree of equilibration, the latter expressed as the relative recovery of the drug *in vitro* ($REC_{in\ vitro}$). Three concentrations are studied in a consecutive and randomised fashion with an adequate washout period between each concentration. The following procedure is applied: the probes are placed in a buffer solution containing the drug of interest at a known concentration (C_m). Constant stirring is applied in the external medium. The experiments are performed at 37 °C in triplicate (3 probes/concentration). The probes are perfused with a 'blank' perfusate solution. Depending on the drug and the analytical sensitivity, 2 to 15 min dialysate fractions are collected (C_d).

Calculations of $REC_{in\ vitro}$ are based on a minimum of 4 samples from each probe at equilibrium. Following the collection period, probes are removed from the drug-containing media, still with constant perfusion, touched with a soft tissue to remove excess fluid from the external surface and placed in 'blank' buffer. Probe perfusion with the 'blank' buffer solution is continued to observe the post-dialysis profile. Such monitoring may indicate the occurrence of drug adhesion to the probe system.

Relative probe recovery ($REC_{in\ vitro}$) is calculated as:

$$REC_{in\ vitro}(\%) = (C_d/C_m) \times 100 \qquad (1)$$

For CMA/12 (3 mm) and CMA/20 (10 mm) probes, respectively, $REC_{in\ vitro}$ averaged $27 \pm 3\%$ and $55 \pm 7\%$ for morphine and $30 \pm 4\%$ and $35 \pm 6\%$ for artemisinin (mean $\pm$ SD); all values being independent of C_m.

In vitro experiments for morphine show a lag-time in probe response of maximally 2 min after placing the probe in a morphine-containing medium (Figure 1). Morphine concentrations rapidly increase to a stable level and rapidly decrease during washout for both probe types.

Perfusion of probes placed in an artemisinin-containing solution followed by perfusion in 'blank' buffer demonstrated a lag-time of 45 to 60 min for equilibration depending on probe type (Figure 2). The post-dialysis profiles indicate a significant adhesion of artemisinin to the MD system components. A lengthy dialysate stabilisation period has also been reported for phenytoin due to solubility problems (Ekblom *et al.*, 1992).

Method of retrodialysis (RD)

The method of RD is calculated by the 'reverse' relative recovery or relative loss (RL) of a substance (Wang *et al.*, 1993; Wong *et al.*, 1993). In contrast to the relative recovery determination, microdialysis probes are immersed in drug-free media ($C_m = 0$) and are perfused with a known drug concentration (C_{in}). *In vivo*, either an internal standard is used during the whole perfusion experiment or the drug of interest is perfused before drug administration is started (Ingvast Larsson, 1991; Scheller and Kolb, 1991; Ståhle, 1991; Van Belle *et al.*, 1993; Wang *et al.*, 1993; Malhotra *et al.*, 1994; Ståhle, 1994).

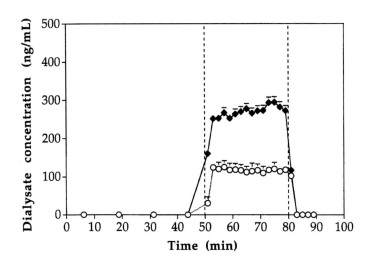

Figure 1 Time aspects of equilibration between probe perfusate and the external medium for morphine with sampling interval (Δt) 2 min for 3 mm CMA/12 (open circles) and 10 mm CMA/20 (filled squares) probes. $C_m = 500$ ng/ml. Data represented as mean values $\pm$ SD ($n = 3$ probes). The dotted line represents the time of probe placement in and removal from morphine-containing medium.

The RL of drug from the perfusion solution is obtained through:

$$RL(\%) = ((C_{in} - C_d)/C_{in}) \times 100 \qquad (2)$$

which is assumed to equal the relative recovery.

For morphine, $REC_{in\ vitro}$ determined by the RD method averaged $28 \pm 2\%$. For artemisinin, it averaged $75 \pm 7\%$ ($n = 3$ probes).

Method of no-net flux (NNF)

NNF, or the 'difference method', is based on determining mass transport of the drug across the microdialysis membrane as a function of perfusate concentration (Lönnroth *et al.*, 1987; Menacherry *et al.*, 1992). NNF requires a constant concentration of the substance of interest in the external medium, whether it is *in vitro* in the buffer or in an *in vivo* situation. The difference between C_d and C_{in} is a measure of the net transport of drug across the membrane. When the concentration of the drug in the perfusate is lower than C_m, substance diffusion occurs from the medium into the probe. The diffusion process is reversed when the concentration is higher in the perfusate. The point of no-net transport is the condition at which $C_{in} = C_m$.

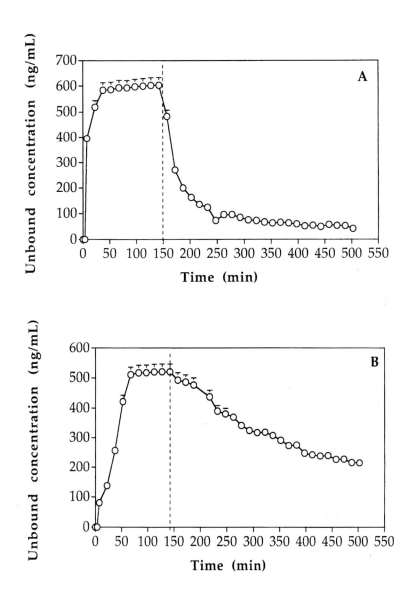

Figure 2 Time aspects of equilibration between probe perfusate and the external medium for artemisinin with sampling interval (Δt) of 15 min for (A) 3 mm CMA/12 probes, and (B) 10 mm CMA/20 probes. Mean values ± SD, $n = 3$. The dotted line represents the time of removal of the probe from the drug-containing medium to 'blank' medium. $C_m = 1500$ ng/ml.

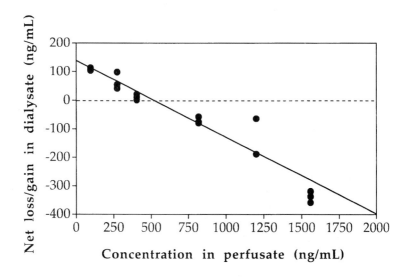

Figure 3 No-net flux experiment of morphine *in vitro* in one probe. The net loss or gain of morphine in dialysate (ΔC) is described for one probe as a function of the perfusate concentration (C_{in}). The intercept with the *x*-axis is 519 ng/ml and the slope (relative recovery) is 27%.

A C_m of 500 ng/ml is employed for both morphine and artemisinin. Six concentrations of C_{in} (100, 300, 400, 900, 1200 and 1500 ng/ml) of both drugs are randomly perfused through the probe and C_d is measured. After each change in perfusate concentration, dialysates are allowed to equilibrate, after which 3 to 4 samples are collected for analysis.

A NNF plot for morphine is shown in Figure 3. The estimated *in vitro* C_m of 550 ± 49 ng/ml in buffer, corresponding to the intercept of the regression line and the *x*-axis, agrees well with the measured morphine C_m of 547 ± 31 ng/ml. The $REC_{in\ vitro}$ equals the slope of the regression line. For morphine NNF gives a $REC_{in\ vitro}$ of 30%.

For artemisinin, the NNF estimated *in vitro* C_m in buffer of 240 ± 38 ng/ml is considerably lower than the measured C_m of 510 ± 25 ng/ml. $REC_{in\ vitro}$ based on slope is 63 ± 6%. In plasma *in vitro*, the estimated C_m is 124 ± 10 ng/ml vs. a measured value of 511 ± 13 ng/ml (unbound concentration determined by equilibrium dialysis). This results in an estimated $REC_{in\ vitro}$ of 103%.

Method of low perfusion rate (LPR)

The method of LPR to determine relative recovery also requires a constant concentration of drug in the surrounding medium. If the perfusion rate is slow enough (or the probe is long enough), there is an equilibration with similar concentrations on both sides of the membrane (Menacherry *et al.*, 1992; Sjöberg *et al.*, 1992). This method is not evaluated further in this chapter, the disadvantage being the requirement of constant concentrations *in vivo* and the lengthy perfusions.

Comparison of recovery methods in vitro for morphine and artemisinin

All three methods for quantifying morphine recovery *in vitro* are comparable. $REC_{in\ vitro}$ are for the 3 mm CMA/12 probe $27 \pm 3\%$, $28 \pm 2\%$ and $30 \pm 3\%$ for traditional relative recovery, RD and NNF, respectively.

Like RD, the method of NNF was found to produce notably high values for artemisinin recovery, with a considerable discrepancy occurring when the different methods were used. $REC_{in\ vitro}$ estimated with RD was 75%, from the NNF plot it was 63% while it was 30% by the traditional method. The discrepancy was even greater for plasma, with an $REC_{in\ vitro}$ of 103% with the NNF method. One of the reasons for these deviations might be due to the adhesion of artemisinin to the probe system.

Probe Characterisation *In Vivo*

RD is chosen for *in vivo* calibration of microdialysis probes prior to the PK experiments due to its advantage in not requiring steady-state drug concentrations and not being too time-consuming. It minimises drug exposure to the body region of interest prior to the systemic drug administration. The disadvantage of the method is that probe behaviour during the experiment is not monitored.

Following equilibration with the surrounding tissue for 60 min, the probes are dialysed for 60 to 75 min with solutions containing the substance of interest, after which they are perfused with blank buffer for 60 to 120 min. To shorten long washout periods, higher flow rates might be employed. However, the effects on tissue homeostasis must be considered (Benveniste and Hüttemeier, 1990; Bungay *et al.*, 1990; Dykstra *et al.*, 1992). Samples collected at the end of the washout period are analysed by HPLC to confirm complete drug removal prior to systemic drug administration.

Relative recovery in vivo for morphine and artemisinin

RD-determined $REC_{in\ vivo}$ values for morphine averaged 6.4% (4.8 to 11%; $n = 13$) in the striatum and 44% (20 to 66%; $n = 13$) in blood. A separate RD experiment illustrated that $REC_{in\ vivo}$ remained constant for 240 min. The $REC_{in\ vivo}$ in brain is approximately 30% of the $REC_{in\ vitro}$.

$REC_{in\ vivo}$ values for artemisinin by the RD method average 84% and >99% for the striatal and muscle/blood probes, respectively. Due to the physicochemical properties of artemisinin, the estimation of relative recovery *in vivo* is unreliable. Therefore, quantification of the unbound concentration *in vivo* is difficult. For the *in vivo* artemisinin data, the traditional *in vitro* method of recovery estimation is chosen for the respective probe type. This could give us a comparison of the relative concentrations of artemisinin in different tissues, assuming the recovery *in vivo* is proportional to the recovery *in vitro*. However, in reality this is seldom the case, as the recovery from one probe type may vary in different tissues.

Protein Binding

MD has proven to be a useful method for estimating protein binding *in vitro* (Dubey *et al.*, 1989; Herrera *et al.*, 1990; Ekblom *et al.*, 1992; Le Quellec *et al.*, 1994). *In vitro* binding of drug in plasma is performed in triplicate at 37 °C at each of three concentrations using both the described MD technique and, if a comparison is desired, in parallel with equilibrium dialysis.

CMA/20 probes are first calibrated *in vitro* as described earlier. MD is performed under constant stirring at 37 °C on aliquots of 2 ml of fresh, non-heparinised plasma. The desired drug concentration in the plasma was achieved with no greater than 5% aqueous solution. Four to six dialysate fractions are collected following a time period required for stabilisation of dialysate concentrations. C_ds measured following perfusion of a plasma aliquot are adjusted for the probe's respective $REC_{in\ vitro}$ value to obtain the fraction unbound in plasma (f_u), as follows:

$$f_u = (C_d/REC_{in\ vitro})/\text{plasma drug concentration} \tag{3}$$

Based on MD and equilibrium dialysis, respectively, artemisinin was determined to be $57 \pm 2\%$ and $71 \pm 4\%$ bound in human plasma and $68 \pm 10\%$ and $56 \pm 3\%$ bound in rat plasma.

Protein binding can also be easily estimated *in vivo* with MD, with the use of the venous probe and blood samples collected in parallel. Plasma is harvested and f_u is calculated according to Equation 3, except that $REC_{in\ vivo}$ is used.

SPECIFIC APPLICATIONS

In this section, the preparation for and the performance of the *in vivo* MD experiments are described. For morphine, PD evaluation is also performed in parallel, with respect to the antinociceptive effect and blood gas status. These measurements are not described further in this chapter, but the technique is described in the references indicated (Ekblom *et al.*, 1993a; Ekblom *et al.*, 1993b; Gårdmark *et al.*, 1993).

Surgery

MD probes are implanted in the striatum, pectoral muscle (artemisinin experiment), and venous blood. Surgery is performed under chloral hydrate anaesthesia (artemisinin experiment) or Hypnorm® (morphine experiment) at least 24 h prior to the start of the experiment. The initial doses are 300 mg/kg i.p. (chloral hydrate) and 0.2 mg/kg i.m. (Hypnorm®) with additional doses of 100 mg/kg chloral hydrate or 0.05 mg/kg Hypnorm® administered as required.

PE-50 cannulae (25 cm length and 0.58 mm i.d.; Intra Medic, Sweden) are inserted into the left femoral artery and vein for the collection of blood samples and for intravenous (i.v.) drug administration, respectively. A 30 cm piece of PE-50 tubing is also looped subcutaneously distal to the posterior surface of the neck to enable the warming of the perfusion solution to body temperature before it enters the striatal probe.

Probe Placement

For the blood probe, a small incision is made in the skin to expose the right external jugular vein. A CMA/20 probe is inserted into the vein with the aid of a plastic i.v. guide cannula through the pectoral muscle to which it is sutured. For muscle MD, a CMA/20 probe is placed in the superficial layer of the left pectoral muscle, also using a plastic guide. Venous probes are perfused with a 0.1% low molecular weight heparin solution prior to *in vivo* placement to minimise the incidence of clotting around the probe membrane. The openings of the inlet and outlet tubing of the probes are sealed.

For implantation of the striatal probe, the rat is placed in a stereotaxic frame (David Kopf Instruments, Tujunga, USA) and a midsagittal incision is made to expose the skull. Guide cannula (CMA/12) placement is accomplished using a micromanipulator mounted on the stereotaxic frame. The stereotaxic coordinates from bregma are: anterior 0.8 mm; lateral, 2.7 mm, and ventral to the dura mater, 3.8 mm (Paxinos and Watson, 1986). The guide cannula is secured to the skull with an anchor screw and dental cement, over which the skin is sutured. The protruding ends of all cannulae and probe tubings are led subcutaneously to the posterior surface of the neck where they are protected by means of a plastic 'cap' sutured to the skin. Rats are then placed in a CMA/120 system for the freely-moving animals to recover for 24 h.

Study Design for *In Vivo* MD

Prior to the experiment, dialysate fractions of 20 to 30 μl are collected over a 60 min period and analysed by HPLC to ensure the absence of analytical interferences. Following this equilibration period, RD is performed as described previously.

Drug is administered i.v. either as a short- or a long-term infusion. Dialysate fractions are collected into pre-weighed vials from all probes at 5 to 30 min intervals during, and for 3 to 8 h after cessation of the infusion. Vials are weighed following fraction collection and the true flow rate is estimated as the difference in pre- and post-collection vial weights divided by the sampling time interval.

To validate the MD technique, blood samples (150 to 500 μl) are collected at specified time intervals and analysed for drug content. Blood samples are centrifuged at 5000 r.p.m. for 5 min, the plasma separated and frozen at $-20\,^{\circ}$C until analysis.

Data Analysis

The unbound *in vivo* tissue concentrations (C_u) of morphine and artemisinin are calculated as the C_d adjusted for the respective probe recovery value. Tissue drug $t_{1/2}$ is calculated from the terminal phase of the disposition curves. Venous $t_{1/2}$s are compared with $t_{1/2}$s determined from the measured plasma concentrations. Total AUC (AUC_{tot}) is obtained from the sum of the products of C_u and the collection time interval (Δt), with the residual area calculated from the last measured concentration (C_n) and the elimination rate constant (λ_n):

$$AUC_{tot} = \sum C_u \cdot \Delta t + C_n / \lambda_n \qquad (4)$$

Tissue drug concentrations and AUCs are expressed relative to each other in order to determine the degree of tissue equilibration. Modelling of PK/PD data is performed using PCNONLIN (Statistical Consultants, 1986).

Results

Morphine

Morphine is found to be rapidly transported into the brain. The time course of morphine concentrations in brain is similar to that in blood for the high dose (40 mg/kg); for the low dose (10 mg/kg) the decline in brain is somewhat slower than in blood (Figure 4). Data from two rats per dose are presented.

Individual AUC ratios brain:blood are 0.08 and 0.26 for the 10 mg/kg dose, and 0.19 and 0.09 for the 40 mg/kg dose of morphine. The variability lies in the AUC in brain. The individual $t_{1/2}$s are 59 min and 54 min in brain and 34 min and 40 min in blood for the 10 mg/kg dose. The corresponding $t_{1/2}$s are 44 min and 42 min in brain and 40 and 35 min in blood for the 40 mg/kg dose.

Artemisinin

Artemisinin demonstrated a high distribution in both brain and muscle. Post-infusion artemisinin concentrations in the striatum declined in parallel with blood concentrations following an initial more rapid decrease (Figure 5); levels in the muscle declined more slowly. The half-lives averaged 2.6 ± 0.9 h, 3.9 ± 1.8 h and 2.4 ± 0.2 h for striatum, muscle, and blood, respectively. For striatal:blood, muscle:blood, and striatal:muscle comparisons, respectively, ratios of artemisinin concentrations at the end of the infusion averaged 1.6 ± 0.6, 1.2 ± 0.5, and 1.4 ± 0.2, whereas ratios were 1.2 ± 0.1, 1.5 ± 0.3, and 0.8 ± 0.2 for AUC_{tot}. However, these concentration-time profiles and PK data should be interpreted in the light of the problems encountered when performing MD on a lipophilic drug like artemisinin.

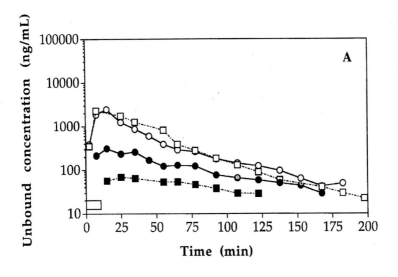

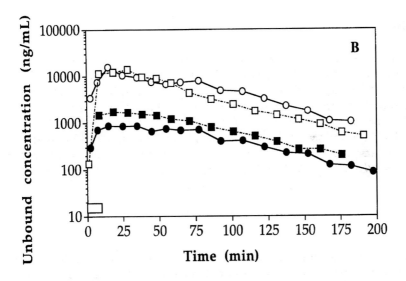

Figure 4 Individual morphine concentration-time profiles in brain (filled symbols) and in blood (open symbols) after (A) a 10 mg/kg dose and (B) a 40 mg/kg dose of morphine hydrochloride as a 10 min infusion.

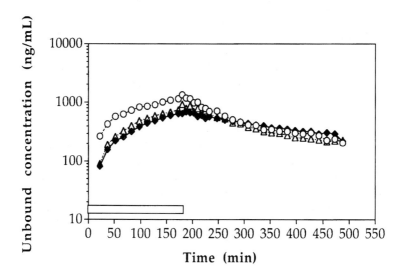

Figure 5 Mean concentration-time profiles of artemisinin in striatum (open circles), muscle (filled diamonds) and venous blood (open triangles) (corrected for recovery *in vitro*) during and following a 180 min infusion (bar) of 24.5 mg/kg/h of artemisinin.

PROBLEM SOURCES AND 'QUALITY CONTROL'

The use of a recovery method *in vivo* is a prerequisite for PK experiments in different tissues, especially when the interest is focused on drug equilibration to the brain. Here, true unbound concentrations in brain ECF and blood are required. RD is chosen as the best method available today. The disadvantage of the RD method as used by us is that when performing a RD experiment of the drug of interest before the start of the PK experiment, probe behaviour during the whole experiment, which might be of value, is not obtained. The use of an internal standard gives this information. However, the disadvantage with using an internal standard is that the transport across the probe *in vivo* might not be the same as that of the drug of interest (Ståhle, 1991). In addition, possible pharmacological effects of an internal standard when performing a PK/PD experiment must be considered.

Experiments with artemisinin illustrated the restricted applicability of the commonly employed methods for estimating the compound's recovery, or of the MD method such as that used for this type of drug. This experience cautions against a generalised application of methods estimating *in vivo* recovery without prior *in vitro* validation for each substance.

In characterising drug kinetic behaviour *in vitro* and *in vivo*, the MD technique is dependent on a rapid equilibration between C_d and C_m. In other words, there should be minimal or no lag-time between kinetic changes occurring around the probe and in corresponding changes in dialysate drug concentrations. Such delay may be due to the adhesion of a compound to the MD probe membrane and to tubings or to temporal aspects of probe membrane passage. The occurrence of such phenomena may question the validity of MD in characterising rapid changes in the regional pharmacokinetics of a drug and may result in erroneous evaluation of PK/PD data. Accordingly, prior characterisation of the compound's equilibration and adhesive behaviour is pivotal in determining its candidacy for MD. It is also important to establish (both *in vivo* and *in vitro*) any dependency of recovery on C_m and flow rate prior to initiating MD *in vivo*.

MD allows for both drug concentration and PD effect to be measured simultaneously in the local environment. The integration of PK and PD measurements with MD are especially suitable for drugs where PD measurements are of a chemical nature, like changes in transmitter levels, but can also be used where the PD measurement is of another type. To date, there are only a few examples in the literature where both concentrations of a drug at the effect site are measured in parallel with the PD measurement (Pettit *et al.*, 1990; Bradberry *et al.*, 1993). In this respect, MD offers a great possibility.

CONCLUDING REMARKS

There is little doubt of the unparalleled potential of MD in the field of PK/PD. To date, the MD technique is not yet fully exploited in parallel PK/PD experiments. The advantages of MD include the possibility to estimate C_u in blood and in tissues in one rat, the possibility of frequent sampling in small animals like rats (or mice) and the decreased number of animals required to obtain qualitative tissue data. The probes can be used up to 3–4 times. The limitations include the problems inherent in the calculation of recovery, although methods such as retrodialysis have overcome much of that problem. Careful *in vitro* determination of drug behaviour in the MD system should precede any *in vivo* studies, as not all drugs are suitable to be studied using MD. Lipophilic drugs with adhesive properties might have special problems.

Collaborators

Artemisinin studies: Dr Michael Ashton, Division of Biopharmaceutics and Pharmacokinetics, Uppsala University.

REFERENCES

Alonso, M.J., Bruelisauer, A., Misslin, P. and Lemaire, M. (1995) Microdialysis sampling to determine the pharmacokinetics of unbound SDZ ICM 567 in blood and brain in awake, freely-moving rats. *Pharm. Res.*, **12**, 291–294.
Benveniste, H. and Huttemeier, P.C. (1990) Microdialysis — theory and application. *Prog. Neurobiol.*, **35**, 195–215.

Bradberry, C.W., Nobiletti, J.B., Elsworth, J.D., Murphy, B., Jatlow, P. and Roth, R.H. (1993) Cocaine and cocaethylene: Microdialysis comparison of brain drug levels and effects on dopamine and serotonin. *J. Neurochem.*, **60**, 1429–1435.

Bungay, P.M., Morrison, P.F. and Dedrick, R.L. (1990) Steady-state theory for quantitative microdialysis of solutes and water *in vivo* and *in vitro*. *Life Sci.*, **46**, 105–119.

Deguchi, Y., Terasaki, T., Ymamada, H. and Tsuiji, A. (1992) An application of microdialysis to drug tissue distribution study: *In vivo* evidence for free-ligand hypothesis and tissue binding of β-lactam antibiotics in interstitial fluids. *J. Pharmacobio-Dyn.*, **15**, 79–89.

Dubey, R.K., C B, Inoue, M. and Wilkinson, G.R. (1989) Plasma binding and transport of diazepam across the blood-brain barrier. No evidence for *in vivo* enhanced dissociation. *J. Clin. Invest.*, **84**, 1155–1159.

Dykstra, K.H., Hsiao, J.K., Morrison, P.F., Bungay, P.M., Mefford, I.N., Scully, M.M. and Dedrick, R.L. (1992) Quantitative examination of tissue concentration profiles associated with microdialysis. *J. Neurochem.*, **58**, 931–940.

Ekblom, M., Hammarlund-Udenaes, M., Lundqvist, T. and Sjöberg, P. (1992) Potential use of microdialysis in pharmacokinetics: A protein binding study. *Pharm. Res.*, **9**, 155–158.

Ekblom, M., Gårdmark, M. and Hammarlund-Udenaes, M. (1993a) Pharmacokinetics and pharmacodynamics of morphine-3-glucuronide in rats and its influence on the antinociceptive effect of morphine. *Biopharm. Drug Dispos.*, **14**, 1–11.

Ekblom, M., Hammarlund-Udenaes, M. and Paalzow, L.K. (1993b) Modeling of tolerance development and rebound effect during different intravenous administration of morphine to rats. *J. Pharmacol. Exp. Ther.*, **266**, 244–252.

Gårdmark, M., Ekblom, M., Bouw, R. and Hammarlund-Udenaes, M. (1993) Quantification of effect delay and acute tolerance development to morphine in the rat. *J. Pharmacol. Exp. Ther.*, **267**, 1061–1067.

Herrera, A.M., Scott, D.O. and Lunte, C.E. (1990) Microdialysis sampling for determination of plasma protein binding of drugs. *Pharm. Res.*, **7**, 1077–1081.

Hurd, Y.L., Kehr, J. and Ungerstedt, U. (1988) *In vivo* microdialysis as a technique to monitor drug transport: Correlation of extracellular cocaine levels and dopamine overflow in the rat brain. *J. Neurochem.*, **51**, 1314–1316.

Ingvast Larsson, C. (1991) The use of an "internal standard" for control of the recovery in microdialysis. *Life Sci.*, **49**, 73–78.

Le Quellec, A., Dupin, S., Tufenkji, A.E., Genissel, P. and Houin, G. (1994) Microdialysis: An alternative for *in vitro* and *in vivo* protein binding studies. *Pharm. Res.*, **11**, 835–838.

Lönnroth, P., Jansson, P.A. and Smith, U. (1987) A microdialysis method allowing characterisation of intercellular water space in humans. *Am. J. Physiol.*, **253**, E228–231.

Lönnroth, P., Carlsten, J., Johson, L. and Smith, U. (1991) Measurements by microdialysis of free tissue concentrations of propranolol. *J. Chromatogr.*, **568**, 419–425.

Malhotra, B.K., Lemaire, M. and Sawchuk, R.J. (1994) Investigation of the distribution of EAB 515 to cortical ECF and CSF in freely moving rats utilising microdialysis. *Pharm. Res.*, **11**, 1223–1232.

Menacherry, S., Hubert, W. and Justice, J.J. (1992) *In vivo* calibration of microdialysis probes for exogenous compounds. *Anal. Chem.*, **64**, 577–583.

Moghaddam, B. and Bunney, B.S. (1989) Ionic composition of microdialysis perfusing solution alters the pharmacological responsiveness and basal outflow of striatal dopamine. *J. Neurochem.*, **53**, 652–654.

Olson, R.J. and Justice, J.J. (1993) Quantitative microdialysis under transient conditions. *Anal. Chem.*, **65**, 1017–1022.

Paxinos, G. and Watson, C. (1986) *The rat brain in stereotaxic coordinates* (2nd ed.), Academic Press, New York.

Pettit, H.O., Pan, H.-T., Parsons, L.H. and Justice, J.B.J. (1990) Extracellular concentrations of cocaine and dopamine are enhanced during chronic cocaine administration. *J. Neurochem.*, **55**, 798–804.

Sarre, S., Delue, D., Van Belle, K., Ebinger, G. and Michotte, Y. (1995) Quantitative microdialysis for studying the *in vivo* L-DOPA kinetics in blood and skeletal muscle of the dog. *Pharm. Res.*, **12**, 746–750.

Scheller, D. and Kolb, J. (1991) The internal reference technique in microdialysis: A practical approach to monitoring dialysis efficiency and to calculating tissue concentration from dialysate samples. *J. Neurosci. Meth.*, **40**, 31–38.

Scott, D.O., Sorenson, L.R. and Lunte, C.E. (1990) *In vivo* microdialysis sampling coupled to liquid chromatography for the study of acetaminophen metabolism. *J. Chromatogr.*, **506**, 461–469.

Sjöberg, P., Olofsson, I.M. and Lundqvist, T. (1992) Validation of different microdialysis methods for the determination of unbound steady-state concentrations of theophylline in blood and brain tissue. *Pharm. Res.*, **9**, 1592–8.

Ståhle, L. (1991) Drug distribution studies with microdialysis: I. Tissue dependent difference in recovery between caffeine and theophylline. *Life Sci.*, **49**, 1835–1842.

Ståhle, L. (1994) Zidovudine and alovudine as cross-wise recovery internal standards in microdialysis experiments? *J. Pharmacol. Toxicol. Meth.*, **31**, 167–169.

Telting-Diaz, M., Scott, D.O. and Lunte, C.E. (1992) Intravenous microdialysis sampling in awake, freely-moving rats. *Anal. Chem.*, **64**, 806–810.

Ungerstedt, U. (1984) Measurement of neurotransmitter release by intracranial dialysis. In C.A. Marsden (Ed.), *Measurement of Neurotransmitter Release In Vivo*, Wiley and Sons, New York, pp. 81–107.

Van Amsterdam, C., Buokhabza, A., Ofner, B., Pacha, W. and Lemaire, M. (1995) Measurement of free concentration of SDZ ICM 567 in blood and muscle using microdialysis sampling. *Biopharm. Drug Dispos.*, **16**, 521–527.

Van Belle, K., Dzeka, T., Sarre, S., Ebinger, G. and Michotte, Y. (1993) *In vitro* and *in vivo* microdialysis calibration for the measurement of carbamazepine and its metabolites in rat brain tissue using the internal reference technique. *J. Neurosci. Meth.*, **49**, 167–173.

Wang, Y., Wong, S.L. and Sawchuk, R.J. (1993) Microdialysis calibration using retrodialysis and zero-net flux: application to a study of the distribution of zidovudine to rabbit cerebrospinal fluid and thalamus. *Pharm. Res.*, **10**, 1411–9.

Wong, S.L., van Belle, K. and Sawchuk, R.J. (1993) Distributional transport kinetics of zidovudine between plasma and brain extracellular fluid/cerebrospinal fluid in the rabbit: Investigation of the inhibitory effect of probenecid utilizing microdialysis. *J. Pharmacol. Exp. Ther.*, **264**, 899–909.

II.2 MICRODIALYSIS MEASUREMENTS OF FREE DRUG CONCENTRATIONS IN BLOOD AND BRAIN

C. VAN AMSTERDAM, P. MISSLIN AND M. LEMAIRE

*Drug Metabolism & Pharmacokinetics, Sandoz Pharma Ltd.,
CH-4002 Basel, Switzerland*

One of the most sensitive and versatile methods to measure drug passage through the blood-brain barrier (BBB) is microdialysis. In this study, microdialysis measurements of SDZ ICM 567, a new 5-HT$_3$ receptor antagonist with a potential efficacy against various central nervous system (CNS) disorders, were carried out in different brain areas and jugular vein of freely moving rats.

INTRODUCTION

The objective of simultaneous blood and brain microdialysis experiments is to characterize the brain disposition of new central nervous system (CNS) active compounds in an early stage of development (Alonso *et al.*, 1995). For this purpose, the test compound is applied intravenously, either as a single bolus or as an infusion. Microdialysis experiments are performed in freely moving animals. The test compound concentrations are generally analysed by means of HPLC, radioimmunoassays (RIA) and LC-MS. Recovery is determined *in vivo* using the 'difference' method (Ståhle, 1991) or the 'reference' method (Wang *et al.*, 1993). A comparison of the concentration-time profiles in brain and blood allows a quantification of the transport mechanisms between the two compartments.

MATERIALS

Rats

For the microdialysis studies male Hanover Wistar rats (No. 139 KF) are used. Care has to be taken of the body weight of rats to be used in brain microdialysis studies, because the stereotaxic coordinates are only reliable for body weight of 270 g to 300 g.

Solutions

Artificial cerebrospinal fluid (CSF) for brain microdialysis (pH 7.4)

5.89 mM	Glucose
126 mM	NaCl
27 mM	$NaHCO_3$
2.4 mM	KCl
0.5 mM	Na_2SO_4
0.5 mM	KH_2PO_4
1.1 mM	$CaCl_2$
0.8 mM	$MgCl_2$

The pH is adjusted by bubbling 95% O_2/5% CO_2 in the solution.

Phosphate buffer for blood microdialysis (pH 7.4)

16 mM	$NaHPO_4 \cdot 2H_2O$
4 mM	KH_2PO_4
100 mM	NaCl

All solutions used for microdialysis are prepared with chemicals of analytical grade and deionized water. They are filtered through a sterile single use filter (FP 030/3, 0.2 μm, Schleicher and Schüll, Dassel, Germany) before use. After filling the syringes, the fluids are degassed in an ultrasonic bath for at least 10 min.

Microdialysis Probes

Brain microdialysis probes

For brain microdialysis in the frontal cortex of rats the CMA 12 probe from CMA Medicine, Sweden, is used. The membrane is of 4 mm length with an O.D. of 0.5 mm. The molecular cut-off is 20,000 Daltons. An intracerebral guide and a dummy probe are delivered together with the probes. It is strongly recommended to use the siliconized guide for easy removal of the dummy probe on the day of the experiment.

Blood microdialysis probes

For blood microdialysis the flexible probe CMA 20 from CMA Medicine, Sweden, is used. The membrane is of 10 mm length with an O.D. of 0.5 mm. The molecular cut-off is 20,000 Daltons. The probe is also available with a 4 mm membrane, but in order to increase recovery, which is directly dependent on the membrane length, it is recommended to use the 10 mm membrane whenever possible.

The CMA 20 is also used for muscle microdialysis in the pectoralis muscle of rats. The flexible probes are delivered with stainless steel introducers (with a plastic handle) and plastic guide tubes. The steel introducer fits perfectly into the plastic tube and goes about 2 mm beyond its tip, so that the sharp point of the introducer is exposed. The guide tube is slit at the handle end to enable easy removal.

Microdialysis System

Perfusion pumps

CMA/100 or the CMA/102 microdialysis pump (CMA Medicine, Sweden), mounted with a 1.0 ml or a 2.5 ml glass syringe are used. The perfusion flow rate in brain and in blood is 2 μl/min for all types of probes.

Connection between the syringe and the probe

In order to avoid twisting and tangling of the perfusion line due to animal movement, a swivel joint with the desired number of channels (normally 2) is put between the syringe and the probe. The connections swivel-syringe and swivel-probe inlet tubing are made with a PE 10 polyethylene tube (Portex Ltd., Hythe, Kent, England) of an appropriate length. It is also necessary to connect the swivel joint to the head of the animal through a flexible cable to absorb the traction provoked by the animal movements. This cable is fixed on the head of the animal to an anchor screw with an alligator clip. On the opposite side the cable is installed into a counterbalanced arm, which is fixed at the cage wall.

The above described system is produced by Alpha M.O.S., Toulouse, France.

Cage

During microdialysis the rat is placed in a plexiglass cage (length 40 cm, width 40 cm, height 60 cm), specifically designed for microdialysis studies with freely moving rats (Alpha M.O.S., Toulouse, France).

Collection of the samples

This is usually done in 200 μl Eppendorff tubes. The tube connected to the outlet of the probes is pushed into a clip fixed to the flexible cable at a short distance to the rat's head.

Analysis

Analysis of parent drug in the microdialysis samples is normally achieved by high performance liquid chromatography (HPLC), e.g. MT2 system, Kontron

Instruments, Zürich, Switzerland. The use of column switching HPLC methods may allow a better separation. The amount of drug is monitored either with UV or fluorescence detection. However, poor UV absorption and fluorescence limit the use of this type of detections. A possible alternative is the use of LC-RID, which requires working with radioactive labelled drugs (Alonso et al., 1995; Everett et al., 1989).

For the quantification of extremely low amounts of drugs in brain dialysate, LC-MS provides an interesting alternative. The combination of a Kontron HPLC-System 400 with a Thermospray/Mass spectrometer TSQ-700 from Finnigan was used successfully to quantify the amount of a drug with poor CNS distribution in brain tissue of rats; such a detection was impossible using classical brain extraction methods. An advantage of this method is the possibility to analyse reference compound (see below) and test compound in one analytical run. Finally, the analysis of the dialysate using RIA represents one of the most convenient techniques when this type of assay is available.

METHODS

Surgery

Anaesthesia for surgery steps is exclusively induced by and maintained with methoxy-flurane (Metofane®, Pitman-Moore) inhalation (induction 3.5% in room air, maintenance 1.5% in room air). The rat is placed in a plexiglass anaesthetic induction chamber, connected to an anaesthetic machine (HNG 4, H. Hölzel, Dorfen, Germany). After induction of anaesthesia, the rat is connected to the anaesthetic machine with a plastic face mask. Rectal temperature is maintained at $37.7 \pm 0.2\,°C$ by means of a thermoregulatory heating pad (CMA/150, Carnegie Medicine, Sweden).

Surgery for microdialysis probes

One week prior to drug administration, the surgical preparations for the brain microdialysis probes are performed. After shaving the head of the anaesthetized rat, the animal is placed in a stereotaxic apparatus (Lab Standard Stereotaxic frame with rat adaptor kit, Stoelting Co., Illinois, USA). Then the head skin is smeared with an antiseptic solution. A crucial step is the introduction of the ear bars into the auditory meatus of the ears. The operator has to be sure that the ear bars are properly inserted, otherwise the rat's head will drop down due to the pressure on the skull while drilling the screwholes. When the ear bars are well positioned, the rat blinks, and the eyes slightly retracts into the orbit. The distance between the ear bars should be approximately 11 mm.

A midline incision of about 2–3 cm is made through the skin and the periost down to the skull is parallel to the sagittal suture. The skull surface is freed from the periost and soft tissue with cotton wool sticks (moistened with saline), bregma is localized and used as the reference point for positioning the guide cannula in the frontal cortex. The guide is attached to the guide holder of the stereotaxic frame. Three holes for fixing screws (3 mm length and 0.5 mm diameter) are drilled in the skull with an electric drill (EWL K9, Kavo, Leutkirch, Germany), two on the left and one on the right part of the skull, each before and behind bregma. In order to achieve suitable fixing in the bone, the diameter of the holes is slightly less than the diameter of the screws.

Starting from zero at the stereotaxic frame, displacements in the anterior-posterior and lateral direction are made until the desired lateral and anterior-posterior coordinates are obtained. The tip of the guide shaft is lowered until it makes contact with the skull, at this point the hole for the guide has to be drilled. This position is marked on the skull with a sharp pencil. The hole (1 mm diameter) is drilled up to the point where the dura mater can be seen. It is highly recommended not to damage the dura mater. The guide is lowered until the tip contacts the dura mater. This is the point zero for the ventral movement, and subsequently the guide is lowered down to the desired region. For appropriate placement of the probe the length of the guide shaft, the length of the dialysis membrane itself, and the size of the area desired for sampling have to be taken into account. For sampling in the frontal right cortex, the following coordinates from bregma are used: anterior +3.2 mm, laterally +1.2 mm and ventrally 1.0 mm.

Once the screws and the guide shaft are in position, all of them are glued together with acrylic cement (Sevriton®, De Trey Dentspley, Konstanz, Germany), applied using a 1 ml plastic syringe. Thereafter the guide holder of the stereotaxic frame is removed and a second layer of acrylic cement is applied onto the skull. At the most anterior part of the system a small metallic shield of 2 cm^2 and at the most posterior part a big screw (15 mm length and 1 mm diameter) are glued within this second layer. The screw is for fixation of the alligator clamp in the microdialysis system. The metallic shield protects the system from being removed by the rat and prevents the guide and the anchor screw from getting caught in the cage. Care has to be taken that the posterior edge of the glue cap is not formed too sharp, otherwise the skin will be injured at this point due to head movements of the animal.

The blood probe is introduced into the jugular vein one day before the experiment. The probe is prepared according to the manufacturer's instructions (soaked 10 min in 70% ethanol and perfused with distilled water, then switching to phosphate buffer) and continuously perfused with phosphate buffer (2 μl/min). In order to prevent blood adhesion to the probe membrane during the time period until the experiment starts, we added low molecular weight heparin (Sigma) to the perfusion fluid (0.1%) for implantation.

The anaesthetized animal is placed on its back on a heating pad. The skin is incised about 2 cm over the pectoralis muscle. The external jugular vein is easily seen, but sometimes a blunt dissection to clear fat and connective tissue is necessary. The pectoral muscle is lifted with toothed forceps and the introducer, covered with the plastic guide tube is inserted into the vein through the muscle tissue at an angle of about 10°. Subsequently the introducer is removed. Blood is coming out of the guide tube if placement is correct. Now the perfused CMA 20 probe is inserted through the guide into the vein and the split ends of the guide tube are torn apart by pulling it upwards and outward. Then the probe is secured to the pectoralis muscle tissue with sutures on both sides of its plastic wings. The inlet and outlet tubes of the probe are tunnelled under the animal's skin to the nape of the neck. For this purpose, an incision is made at the desired point and a cannula is pushed under the skin to the probe implant side. Until this point of the surgery the probe is continuously perfused. Now the perfusion is stopped and the tubes are melted to close. The probe tubes are fixed into the cannula, which is pushed back to exteriorize the probe tubes. Then the skin incisions are closed with tissue staples and the probe tubes are secured with tape to prevent them for slipping under the skin.

Surgery for drug application

For i.v. application of drugs the right femoral vein and for blood sampling the femoral artery of the rat are cannulated with PE 50 following the usual surgery procedures. This is carried out during the surgery for the insertion of the jugular vein probe. The tubes are exteriorized at the neck of the animal and melted to close after filling them with heparin/saline in order to avoid blood clotting.

For drug infusion, the rat is placed in a special in-house designed metabolic cage. The femoral vein catheter is connected to a swivel joint and a syringe mounted in a Harvard pump. The tubings are protected from gnawing by the rat with a flexible metallic cable. For i.v. bolus application, the drug solution is applied through the femoral vein catheter, followed by a 0.3 ml saline flush in order to apply the whole volume.

Microdialysis Experiments

The first step on the day of the experiment is the preparation of the microdialysis probe which is soaked in 70% ethanol for 15 min and then flushed with distilled water. During the flushing the probe is checked for leaks. The syringes (one for brain and one for blood probe perfusion), filled with the degassed perfusion fluids, are now installed into the pump and connected to the proximal extremity of the perfusion circuit of the system. After starting the pumps (flow rate about 20 μl/min), the system is filled with fluid. To flush out any particles and air bubbles present in the system the liquid is allowed to flow freely for at least 10 min. During this period the system is checked carefully for air bubbles and leaks. Then the brain probe is connected to the tubing coming from the swivel joint and placed in an Eppendorff vial filled with artificial CSF, while being perfused with artificial CSF at a rate of 2 μl/min. Subsequently the rat is placed into the cage and connected with the anchor screw on the skull to the alligator clip. The dummy of the intracerebral guide is removed. The probe is introduced carefully into the guide so as not to damage the sensible membrane. It is very helpful to perform this part with two persons, one of them fixing the rat and the other introducing the probe. A short piece of the inlet of the blood probe is cut off open and the probe is then connected to tubing coming from the swivel joint. All connections in the system are made with tubing connection adapters (Carnegie Medicine, Sweden). During the course of perfusion, the animal is allowed to move within the confines of the cage, and is provided with food and water.

The equilibration time after introduction and connection of the rat is usually about 60–90 min. The time is necessary because the insertion of the probe in the brain could produce certain damages. It is recommended to check the flow of liquid at the outlet of the probe during this period by quantification of collected volume/unit of time. Thereafter the drug is administered.

The collection interval of dialysate is highly dependent on the sensitivity of the analytical system. It has to be taken into account that microdialysis only samples the free drug in the tissue examined. Moreover, the amount of drug in the dialysate is reduced due to the fact that the recovery of the probe is not 100%. If possible, the dialysate is collected every 15 min in order to obtain continuous measurements. For brain microdialysis the sampling interval is normally 30 min. With a flow rate of 2 μl/min the sample volume is 60 μl. However, if a sufficient amount of drug for analysis is expected in the dialysate, it is possible to increase the perfusion flow and to decrease the sample interval time.

Estimation of Recovery

Recovery is defined as the ratio of the amount of drug measured in the dialysate to that in the medium surrounding the probe membrane and is estimated either by the reference or the difference method.

Reference method (Retrodialysis)

Utilizing this technique, microdialysis in blood and brain is performed as described above. The only difference is the addition of a reference compound to the perfusion fluid. This technique is described elsewhere (Wang *et al.*, 1993). Briefly, it is assumed that the 'loss' of this reference compound from the probe into the surrounding tissue during microdialysis is equivalent to the normal gain recovery of the test compound from the extracellular space into the probe. The loss is estimated according to the following equation

$$\text{Loss}_{\text{reference compound}} = \text{Recovery}_{\text{test compound}} = 1 - \frac{(C_{\text{out}})}{(C_{\text{in}})} \qquad (1)$$

where (C_{in}) is the concentration of the reference compound in the perfusate (liquid entering the probe) and (C_{out}) is that in the dialysate (liquid leaving the probe), respectively. Hence, the recovery of a given probe can be estimated over each dialysis interval, and the concentration of the test compound in the sampled tissue (C_{tissue}) can be calculated as

$$(C_{\text{tissue}}) = \frac{(C_{\text{dialysate}})}{\text{Recovery}} \qquad (2)$$

where $(C_{\text{dialysate}})$ is the concentration of the drug in the dialysate and recovery is the recovery obtained according to Equation 1.

It is mandatory for this method that the reference compound exhibits similar diffusion kinetics through the microdialysis probe membrane as the test compound. Thus, this compound should be as chemically similar to the test compound as possible. Usually, in the pharmaceutical industry several drugs of a chemical group are synthesized, providing a lot of appropriate candidates. It is also possible to use a radioactive labelled derivative of the test compound. For instance, to examine the distribution of SDZ ICM 567, a new 5-HT$_3$ receptor antagonist, 50 ng/ml [^{14}C]SDZ ICM 567 was added to the perfusate to serve as a reference (van Amsterdam et al., 1995). Thus, (C_{in}) was 50 ng/ml. Assuming that (C_{out}) is 30 ng/ml and the amount of unlabelled drug in the dialysate is 100 ng/ml, according to the given equations the loss and the recovery is 0.4 and the concentration of free SDZ ICM 567 in the sampled tissue is 250 ng/ml.

To check whether both compounds show the same diffusion kinetics, it is recommended to compare the loss of the chosen reference and the gain of the test compound in vitro, e.g. as described in (Wang et al., 1993). Although the same behaviour in vitro does not ensure that the compounds will exhibit identical diffusion kinetics in vivo, this experiment will at least give some indication. However, the selection of a reference is also dependent on the analytical system. It would be advantageous, if the amount of reference and that of the test compound in the dialysate can be assayed simultaneously in one step, and thus avoiding an additional analytical procedure.

Difference method

Using this method, steady state conditions should be established in the animal, e.g. with an i.v. infusion over 16–18 h (Alonso et al., 1995). Then, the implanted probes are perfused with solutions containing 3–4 different concentrations of the compound, e.g. 10, 25, 50 and 100 ng/ml (Alonso et al., 1995). This is realised by means of a liquid switching system (CMA/120, Carnegie Medicine, Sweden). The principle of this method is to estimate the point when no-net mass transport over the probe membrane occurs. The point of no-net flux is that at which perfusate and sampled medium concentrations are identical. It can be determined by linear regression of the net transport for different perfused concentrations. The difference between the concentration of perfusate and dialysate, ΔC, is measured and plotted against the concentration of the perfusate (C_{in}). The slope of the regression line is the recovery.

Normally, the recovery estimated with the difference method is obtained from an appropriate number of animals, e.g. 4 (Alonso et al., 1995). Then, the mean value of the recovery of the probes in brain and blood is used as estimation for the probes used in further microdialysis studies, assuming no significant difference in recovery of the different probes.

Pharmacokinetic Evaluation

The concentration time profiles are individually fitted by computerized methods (Alonso *et al.*, 1995). Pharmacokinetic evaluation of data obtained from microdialysis studies is made by comparing the area under the concentration-time curve (AUC) of the different tissues, e.g. in brain and in blood. In contrast to blood sampling at definite time points, microdialysis samples contain the mean concentration of a period of time (Ståhle, 1992). Thus, the technique itself is an integrating method and the AUC is calculated by summing up the products between sampling interval Δt and the sample concentration Cn (corrected for recovery) with the addition of the remaining area (Cn/β):

$$\text{AUC} = \sum Cn\Delta t + \frac{Cn}{\beta} \qquad (3)$$

RESULTS

Some results from recently performed studies will illustrate the use of this technique in our laboratory. Microdialysis in the frontal cortex and jugular vein of freely moving rats was performed to examine the free, unbound concentrations of SDZ ICM 567, a new 5-HT$_3$ receptor antagonist with potential efficacy against various CNS disorders, after a 10 mg/kg intravenous dose (Alonso *et al.*, 1995). The concentration-vs.-time profile is shown in Figure 1. The AUC$_{0-5h}$ corresponding to the unbound drug in blood was 462 ± 142 ng·ml^{-1}·h (mean $\pm$ SD, $n = 4$). The free unbound amount of

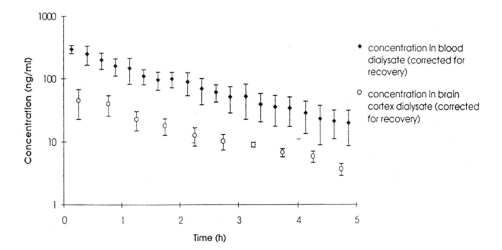

Figure 1 Free concentration-vs.-time profile for SDZ ICM 567 in blood and brain cortex of rats (mean $\pm$ SD of 4 rats) after an i.v. bolus of 10 mg/kg.

the drug (AUC_{0-5h}) determined in the frontal cortex was 86 ± 24 ng·ml^{-1}·h. Thus, the free concentration of SDZ ICM 567 in brain represents approximately 15% of that in blood. It was concluded, that several mechanisms may explain the difference such as active transport out of the brain or some uptake and metabolism processes in brain cells.

Microdialysis was also used to estimate the concentration of SDZ ICM 567 after a 1.5 mg/kg intravenous dose in the blood and the pectoralis muscle of anaesthetized rats (van Amsterdam et al., 1995). In contrast to the results obtained in the brain, the $AUC_{0-2.5h}$ in muscle (168.4 ± 29.7 ng·ml^{-1}·h, mean $\pm$ SEM, $n = 4$) was similar to that in blood (151.6 ± 25.7 ng·ml^{-1}·h). The muscle/blood concentration ratio was close to unity at each individual sampling point. Thus, the distribution of free SDZ ICM 567 into muscle extracellular tissue fluid is best explained by a passive diffusion process.

The recovery of probes for SDZ ICM 567 used in the brain/blood microdialysis study was estimated with the 'difference' method. The recoveries in the blood and in the brain were $46.4 \pm 3.4\%$ and $47.7 \pm 4.4\%$ ($n = 4$ probes, mean $\pm$ SD), respectively. Additional experiments (unpublished data) in freely moving rats have shown that the recovery of jugular vein probes estimated with the addition of 50 ng/ml [^{14}C]SDZ ICM 567 to the perfusion fluid as a reference compound was $49.9 \pm 5.9\%$ ($n = 9$ probes), suggesting identical results for the difference and the reference method.

CONCLUDING REMARKS

Microdialysis represents actually one of the most sensitive and versatile methods to quantify the passage of drugs through the BBB. However, two main problems still remain: the bioanalytical aspect and the estimation of recovery. The bioanalytical aspect includes contamination and detection problems. Whereas efforts should be done to answer the first question, the use of new analytical techniques (capillary electrophoresis, micro HPLC, LC-MS) should improve the use of microdialysis. It is now evident, that in vivo estimations of recovery have to be done and that the mode of analysis has to be adapted to the test compound.

Our actual studies concern simultaneous measurements in different brain areas, e.g. brain cortex and cisterna magna, and in the blood. The technique also allows the use of multiple probes in one rat, and the recovery during the experiment. The first results of these simultaneous brain microdialysis procedures are promising, and this approach should represent one of the next steps in the development of brain microdialysis.

REFERENCES

Alonso, M.J., Bruelisauer, A., Misslin, P. and Lemaire, M. (1995) Microdialysis sampling to determine the pharmacokinetics of unbound SDZ ICM 567 in blood and brain in awake, freely moving rats. Pharm. Res., 12, 291–294.

Everett, D.W., Foley, J.E., Singhvi, S.M. and Weinstein, S.H. (1989) High performance liquid chromatographic method for the radiometric determination of [^{14}C]bucromarone in human plasma utilizing non-radiolabelled bucromarone as an internal standard. *J. Chromatogr.*, **487**, 365–373.

Ståhle, L. (1991) The use of microdialysis in pharmacokinetics and pharmacodynamics. In: Robinson, T. and Justice, T. (Eds.), *Microdialysis in Neurosciences*, pp. 155–174. Elsevier Science Publishers, Amsterdam.

Ståhle, L. (1992) Pharmacokinetic estimations from microdialysis data. *Eur. J. Clin. Pharmacol.*, **43**, 289–294.

Van Amsterdam, C., Boukhabza, A., Ofner, B., Pacha, W. and Lemaire, M. (1995) Measurement of free concentration of SDZ ICM 567 in blood and muscle using microdialysis sampling. *Biopharm. Drug Dispos.*, **16**, 521–527.

Wang, Y., Wong, S. and Sawchuk, R.J. (1993) Microdialysis calibration using retrodialysis and zero-net-flux: Application to a study of the distribution of ziduvidine to rabbit cerebrospinal fluid and thalamus. *Pharm. Res.*, **10**, 1411–1419.

II.3 TRANSCEREBRAL MICRODIALYSIS OF MORPHINE AND MORPHINE 6-GLUCURONIDE

P. SANDOUK,[1,2] M. BARJAVEL,[1] F. STAIN[1] AND J.M. SCHERRMANN[1,2]

[1] *INSERM Unité 26, Hôpital Fernand Widal, Paris, and*
[2] *Laboratoire de Biopharmacie et de Pharmacocinétique,*
Université René Descartes-Paris V, Faculté des Sciences
Pharmaceutiques et Biologiques, Paris, France

Morphine (M), and morphine 6-glucuronide (M6G) were subcutaneously administered at 10 mg/kg in two groups of six freely-moving rats. A transverse microdialysis probe was implanted in the brain cortex and dialysates were collected every 30 min for a period of 4 h. Dialysates were measured by two opiate radioimmunoassays (RIA). Maximum brain opiate concentrations, 41 ± 10 ng/ml (M), 177 ± 43 ng/ml (M6G), were reached at the same T_{max}, 0.75 h, and elimination half-lives ranged from 0.99 to 0.81 h for the 2 compounds. Kinetic parameters confirmed that penetration and elimination rates in the extracellular space of the rat brain cortex for the hydrophilic morphine metabolite was similar to those of morphine.

INTRODUCTION

Several mammalian species excrete morphine 6-glucuronide (M6G) as a minor metabolite and morphine 3-glucuronide (M3G) as a major metabolite of morphine (M). M6G has been shown to exert much stronger analgesic activity than morphine itself after intracerebroventricular (i.c.v.) administration (Paul *et al.*, 1989), whereas M3G has been proposed to act as an antagonist (Smith *et al.*, 1990). It is currently believed that polar metabolites such as glucuronides cannot penetrate into the brain through the blood-brain-barrier by diffusion processes that require lipophilicity (Oldendorf, 1974). Nevertheless, although M6G and M3G are highly polar conjugates, brain morphine glucuronide concentrations have been described in rats and humans after a peripheral morphine administration (Yoshimura *et al.*, 1973; Poulain *et al.*, 1990).

This chapter describes the brain input and output rates of M, and M6G after subcutaneous (s.c.) administration in the conscious rat using the *in vivo* technique of transcortical microdialysis.

MATERIALS AND METHODS

In Vivo Microdialysis

Morphine hydrochloride, and M6G were from Francopia, Paris. The microdialysis probe was an acrylonitrile-sodium methallyl sulphonate membrane (AN69 HF

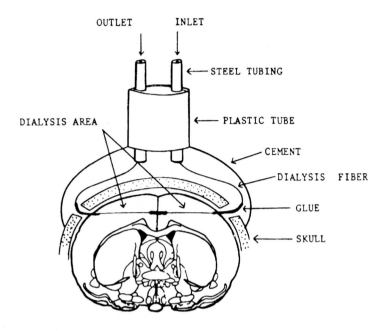

Figure 1 Schematic diagram of a transcortical dialysis preparation in the rat.

Filtral 12; inner diameter (damp) 240 μm, outer diameter 290 μm; molecular weight cutoff 44,000; Hospal Industrie S.A., 69883 Meyzieu, France). The dialysis tubing was coated with epoxy glue, except for a 8 mm dialysis tip corresponding to the length of the rat parietal cortex. The procedure used for the insertion of the transcortical dialysis fibre was essentially the same as that described by Imperato and Di Chiara (1985) except that, in the present study the fibre was implanted in the cortex. Male Sprague Dawley rats (250–300 g, Charles River, France) were anaesthetized with chloral hydrate (300 mg/kg i.p.) and placed in a stereotaxic frame (David Korpf Instruments™, Roucaire, Velizy Villacoublay, France). To implant the transcortical probe, two holes were drilled on the side of the skull (coordinates according to Paxinos and Watson, 1986: A-P, 0.3 mm, M-L 2.0 mm). The dialysis fiber, held straight by an internal tungsten wire was implanted transversely in the brain cortex. The tungsten wire was withdrawn and two stainless steel tubes (23 gauge needle/10 mm long) were glued to the end of the fibre to form the outlet. The probe was fixed to the skull with acrylic dental cement. A schematic diagram of the whole system is shown in Figure 1. After surgery, rats were allowed to recover overnight with food and water available *ad libitum*.

To test the integrity of the blood-brain barrier (BBB) 24 h after probe implantation, labelled α-aminoisobutyric acid (AIB) a neutral amino acid ($M_W = 104$) that does not cross the BBB (Blasberg *et al.*, 1983) was used. Twenty-five μCi of [^{3}H]AIB (25 Ci/mmol, DuPont NEN®, Les Ulis, France) were injected intravenously as a bolus in 3 control rats and 3 other rats 24 h after the dialysis fibre implantation. Serial blood samples were drawn from the femoral artery. The animals were decapitated at the end of the experiment, 10 min after the injection of the radiotracer. The brains were quickly removed and dissected into several regions. Arterial plasma and weighed regions of the brain were prepared for counting the radioactivity, then the unidirectional blood-to-brain transfer constant K_i (Ohno *et al.*, 1978) was calculated as described by Ziylan *et al.* (1988), in control and dialysed rats, for [^{3}H]AIB.

Dialysis experiments were performed 24 h after the probe implantation by perfusing the dialysis probe with Ringer's solution (NaCl 147 mM, KCl 4 mM, CaCl$_2$ 2.4 mM, pH adjusted to 7.3 with NaOH) at a constant flow rate (3 μl/min), using a microinfusion pump (Precidor®, Infors AG, Basel, Swiss). Basal levels were collected and 2 groups of 6 rats received M and M6G respectively at 10 mg/kg (expressed as base) in saline, subcutaneously, in the neck area. The dialysate samples were continuously collected and fractionated every 30 min for 4 h in small vials and kept frozen until assays were performed.

Assays

Morphine was measured by a radioimmunoassay (RIA) using antibodies raised in goats against *N*-carboxymethylnormorphine linked to bovine serum albumin (BSA). Antibodies were specific to morphine and exhibited no cross-reactivity with M3G, M6G (<0.2%) and opiate peptides (<0.05%) as previously described (Sandouk *et al.*, 1991). M6G was measured by a second RIA using an antiserum raised in rabbits with 6-hemisuccinylmorphine linked to BSA as an immunogen. M6G exhibited 100% of cross-reactivity with this antiserum. M or M6G standards in phosphate buffer were used to establish the corresponding opiate standard curves. Limits of quantification were 0.1, and 0.39 ng/ml, respectively. The day-to-day and within-day variations ranged from 6.6 to 8.2% and 1.2 to 5.0% respectively at concentrations ranging from 0.1 to 50 ng/ml for the two RIAs.

In Vitro Recovery of Morphine and Morphine Metabolite through the Dialysis Probe

In vitro recovery of M and M6G through the dialysis membrane was determined using 6 different probes in a Ringer's solution (37 °C) containing from 20 to 200 ng/ml of each opiate. Probes were perfused with Ringer's solution at a flow rate of 3 μl/min. Perfusates were collected and assayed as described above. Recovery was calculated by dividing dialysate opiate concentrations by those in Ringer's solution and expressed as a percentage.

Pharmacokinetics

Concentration-time curves were fitted using the mid-point of the sampling interval (Ståhle, 1992). Maximal opiate concentration (C_{max}) and time (T_{max}) were observed and the area under the experimental curve ($AUC/_0^4$) and terminal half-life were calculated using the trapezoidal and linear regression methods respectively, with the Siphar program (Simed, Créteil, France). Data are expressed as mean ± SEM.

Statistical Analysis

Student's t-test for two independent groups and one-way analysis of variance (ANOVA) were used to compare means of K_i values in various brain regions between control and dialysed rats. Significance was taken as $p < 0.05$. Statistical comparisons for the pharmacokinetic parameters between the two groups of rats (M and M6G) were performed using the Mann-Whitney U-test (unpaired test). The two-tailed p-value <0.05 was taken as the level of significance (GraphPad InStat software, San Diego, CA, USA).

RESULTS

In order to evaluate the damage to the BBB due to implantation of the dialysis fibre in the cortex of rats 24 h later, the unidirectional blood-to-brain transfer constant (K_i) for [³H]AIB was calculated in various brain regions of dialysed rats vs. those of control rats. No significant differences ($p > 0.2$) for all brain regions were found between the two groups of rats (Table 1).

In vitro M and M6G recoveries were respectively 28–33% and 20–32% with external probe concentration ranging from 20 to 200 ng/ml. The time course of extracellular M and M6G in the cortex was similar (Figure 2). The compounds were detected in the sample of the first 30 min and peak levels were reached 0.75 h after the drug administration (Table 2), which then declined in the disposition phase with similar half-lives ($p > 0.05$). Therefore, the main difference was a significantly lower C_{max} and AUC for M compared to M6G ($p < 0.05$). M6G dialysates were also assayed with the morphine-specific RIA. Within the range of M6G concentrations, no morphine was detected (>0.1 ng/ml).

TABLE 1 Regional K_i values for [³H]AIB in control rats and in rats 24 h after the implantation of the transcortical probe (mean ± SEM, $n = 3$). **A** = above the probe. **U** = under the probe.

Brain regions	K_i $(ml \cdot g^{-1} min^{-1} \times 10^3)$	
	Control rats	*Dialysed rats*
Right parietal cortex **A**	2.28 ± 0.11	2.31 ± 0.25
Left parietal cortex **A**	2.26 ± 0.08	2.27 ± 0.19
Right parietal cortex **U**	2.30 ± 0.13	2.15 ± 0.31
Left parietal cortex **U**	2.11 ± 0.10	2.20 ± 0.27
Right frontal cortex	1.97 ± 0.12	2.10 ± 0.19
Left frontal cortex	2.02 ± 0.15	1.99 ± 0.25

TABLE 2 Pharmacokinetic parameters calculated for morphine and morphine 6-glucuronide after determination of concentrations in the dialysates. Data are mean $\pm$ SEM ($n = 6$).

	C_{max} (ng/ml)	T_{max} (h)	AUC_0^A (ng/ml·h^{-1})	$t_{1/2ter}$ (h)
MORPHINE	41*	0.75	62*	0.99
	±10		±11	±0.16
M6G	177	0.75	256	0.81
	±43		±45	±0.14

*M values significantly different at $p < 0.05$ than values for M6G.

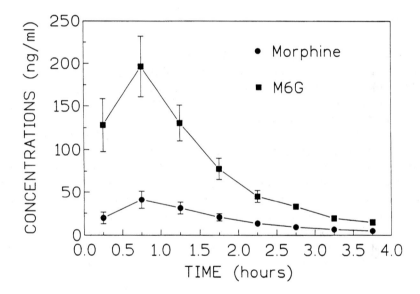

Figure 2 Time-concentration plots of morphine and morphine 6-glucuronide in dialysates. Probes were implanted transversely in the brain cortex. Drugs were injected at a dose of 10 mg/kg s.c. The curves are fitted using the mid-point of the sampling interval (30 min). Data are mean $\pm$ SEM (bars) values ($n = 6$).

DISCUSSION

Glucuronidation is known to be a detoxification pathway for xenobiotics. However, some glucuronides are reported to exert toxic or pharmacological activity (Kroemer et al., 1992). For example, M6G and M3G, the main morphine metabolites, are active brain compounds (Paul et al., 1989; Smith et al., 1990).The hydrophilic property of glucuronides should preclude their brain penetration according to the partition coefficient theory (Oldendorf, 1974). In fact, since the 1970s, the penetration of M6G and M3G in rat brain has been demonstrated after a systemic administration of M6G and M3G (Shimomura *et al.*, 1971; Yoshimura *et al.*, 1973), and more recently the presence of M6G in CSF after oral M administration in humans was reported by Poulain *et al.* (1990).

After an oral administration of M in humans, the mean plasma AUC ratio of M3G to M was 56:1 and for M6G to M, the ratio was 10:1 (Osborne *et al.*, 1990). Moreover, as M3G and M6G are rapidly formed, a sufficient concentration gradient of circulating blood opiates occurs to allow diffusion of M3G and M6G through the BBB. To evaluate the penetration of M6G into the brain, we have used an *in vivo* microdialysis technique which allows the measurement of free compounds in the extracellular fluid of the parietal cortex from rat brain. Because of the continuous collection of dialysates, the rate of drug input and output can be recorded in the same animal.

The validity of the above assumption requires the integrity of the BBB when drugs are administered, 24 h after the implantation of the dialysis fibre in the rat cortex. As clearly seen in Table 1, no significant quantities of AIB can be found in tissue surrounding the probe in dialysed rats compared to control rats. A few hours after probe implantation the BBB restores its functions, as has been demonstrated by several authors using labelled drugs (Benveniste *et al.*, 1984; Terasaki *et al.*, 1992) or by histological examinations (Consolo *et al.*, 1987; L'Heureux *et al.*, 1986). Moreover, drug kinetics in the brain can be evaluated because the *in vitro* recoveries are similar for M and M6G, in the concentration range observed *in vivo*. However, drug measurements through the dialysis membrane only provide a relative estimate of concentrations in the extracellular medium. Therefore the *in vivo* recovery of M in different regions of the rat brain will be lower than the *in vitro* recovery of M from Ringer's solution. For this reason the bioavailable fraction of M and M6G was not investigated in this study.

Our data clearly indicate that morphine 6-glucuronide reached the cortex area as quickly as morphine and declined with parallel kinetics. In accord with previous results, when morphine microdialysis was performed, brain morphine concentrations were maximal 45–60 min after intraperitoneal administration, and decreased to non-detectable values after 3.5 h (Matos *et al.*, 1992). In addition, we demonstrate that M6G has uptake and decline rates in the cortex area that are similar to those of morphine. The unexpected pharmacokinetic behaviour of this hydrophilic conjugate may be explained by *in vitro* experiments (Carrupt *et al.*, 1991); the two morphine glucuronides can exist in conformational equilibrium between hydrophilic and lipophilic forms, depending on the surrounding media. In the vicinity of the BBB, both glucuronides are as lipophilic as M, and could penetrate into the brain with the same kinetics as M.

The use of the morphine-specific RIA demonstrated that no free morphine was present in the extracellular cortex fluid after M6G administration and confirms the chemical stability of this drug, as previously described (Shimomura *et al.*, 1971).

CONCLUDING REMARKS

Using *in vivo* technique of transcortical microdialysis, the input and output rates of morphine and morphine-6-glucuronide were measured following a subcutaneous administration of these drugs in the conscious, freely-moving rats. Our data show that the morphine glucuronide penetrates into and disappears from the brain with kinetics similar to those of morphine.

REFERENCES

Benveniste, H., Drejer, J., Schousboe, A. and Diemer, N.H. (1984) Elevation of the extracellular concentrations of glutamate and aspartate in rat hippocampus during transient cerebral ischemia monitored by intracerebral microdialysis. *J. Neurochem.*, **43**(5), 1369–1374.

Blasberg, R.G., Fenstermacher, J.D. and Patlak, C.S. (1983) Transport of α-aminoisobutyric acid across brain capillary and cellular membrane. *J. Cereb. Blood Flow Metabol.*, **3**(1), 8–31.

Carrupt, P.A., Testa, B., Bechalany A., El Tayar, N., Descas, P. and Perrissoud, D. (1991) Morphine 6-glucuronide and morphine 3-glucuronide as molecular chameleons with unexpected lipophilicity. *J. Med. Chem.*, **34**, 1272–1275.

Consolo, S., Wu, C.F., Fiorentini, F., Ladinsky, H. and Vezzani, A. (1987) Determination of endogenous acetylcholine release in freely moving rats by trans-striatal dialysis coupled to radioenzymatic assay: Effects of drugs. *J. Neurochem.*, **48**(5), 1459–1465.

Imperato, A. and Di Chiara, G. (1985) Dopamine release and metabolism in awake rats after systemic neuroleptics as studied by trans-striatal dialysis. *J. Neurosci.*, **5**(2), 297–306.

Kroemer, H.K. and Klotz, U. (1992) Glucuronidation of drugs: A re-evaluation of the pharmacological significance of the conjugates and modulating factors. *Clin. Pharmacokinet.*, **23**, 292–310.

L'Heureux, R., Dennis, T., Curet, O. and Scatton, B. (1986) Measurement of endogenous noradrenaline release in the rat cerebral cortex *in vivo* by transcortical dialysis: Effects of drugs affecting noradrenergic transmission. *J. Neurochem.*, **46**(6), 1794–1801.

Matos, F.F., Rollema, H. and Basbaum, A.I. (1992) Simultaneous measurement of extracellular morphine and serotonin in brain tissue and CSF by microdialysis in awake rats. *J. Neurochem.*, **58**, 1773–1781.

Ohno, K., Pettigrew, K.O. and Papoport, S.T. (1978) Lower limits of cerebrovascular permeability to nonelectrolytes in the conscious rat. *Am. J. Physiol.*, **253**, H299–H307.

Oldendorf, W.H. (1974) Lipid solubility and drug penetration of the blood-brain-barrier. *Proc. Soc. Exp. Biol. Med.*, **147**, 813–816.

Osborne, R., Joel, S., Trew, D. and Slevin, M. (1990) Morphine and metabolite behavior after different routes of morphine administration: Demonstration of the importance of the active metabolite morphine 6-glucuronide. *Clin. Pharmacol. Ther.*, **47**, 12–19.

Paul, D., Standifer, K.M., Inturrisi, C.E. and Pasternak, G.W. (1989) Pharmacological characterization of morphine-6β-glucuronide, a very potent morphine metabolite. *J. Pharmacol. Exp. Ther.*, **251**, 477–483.

Paxinos, G. and Watson, C. (1986) *The Rat Brain in Stereotaxis Coordinates* (2nd edition). Academic Press, New York.

Poulain, P., Moran Ribon, A., Hanks, G.W., Hoskin, P.J., Aherne, G.W. and Chapman, D.J. (1990) CSF concentrations of morphine 6-glucuronide after oral administration of morphine. *Pain*, **41**, 115–116.

Sandouk, P., Serrie, A., Scherrmann, J.M., Langlade, A. and Bourre J.M. (1991) Presence of morphine metabolites in human cerebrospinal fluid after intracerebroventricular administration of morphine. *Eur. J. Drug Metab. Pharmacokinet.*, Special Issue no. **III**, 166–172.

Shimomura, K., Kamata, O., Ueki, S., Ida, S., Oguri, K., Yoshimura, H. and Tsukamoto, H. (1971) Analgesic effect of morphine glucuronides. *Tohoku J. Exp. Med.*, **105**, 45–52.

Smith, M.T., Watt, J.A. and Cramond, T. (1990) Morphine 3-glucuronide: A potent antagonist of morphine analgesia. *Life Science*, **47**, 579–585.

Ståhle, L. (1992) Pharmacokinetic estimations from microdialysis data. *Eur. J. Clin. Pharmacol.*, **43**, 289–294.

Terasaki, T., Deguchi, Y., Kasama, W., Pardridge, W. and Tsuji, A. (1992) Determination of *in vivo* steady state unbound drug concentration in the brain interstitial fluid by microdialysis. *Int. J. Pharmaceut.*, **81**, 143–152.

Yoshimura, H., Ida, S., Oguri, K. and Tsukamoto H. (1973) Biochemical basis for analgesic activity of morphine-6-glucuronide. I. Penetration of morphine-6-glucuronide in the brain of rats. *Biochem. Pharmacol.*, **22**, 1423–1430.

Ziylan, Y.Z., Lefauconnier, J.M., Bernard, G. and Bourre, J.M. (1988) Effect of dexamethasone on transport of α-aminoisobutyric acid and sucrose across the blood-brain barrier. *J. Neurochem.*, **51**(5), 1338–1342.

II.4 EFFECT OF PERFUSATE TONICITY AND TEMPERATURE ON MICRODIALYSATE AUC VALUES OF ACETAMINOPHEN AND ATENOLOL OBTAINED FROM RAT CORTICAL BRAIN

E.C.M. de LANGE, M. DANHOF, A.G. de BOER AND D.D. BREIMER

Leiden/Amsterdam Center for Drug Research, Division of Pharmacology, Sylvius Laboratories, P.O. Box 9503, 2300 RA Leiden, The Netherlands

This chapter describes an intracerebral microdialysis technique using transversal cellulose acetate microdialysis probe implanted in the cortical brain of male Wistar-derived rats, to measure the effect of tonicity and temperature of the perfusate on the brain microdialysate concentrations of atenolol and acetaminophen. The concentration-time profiles of either atenolol or acetaminophen in the dialysate were determined after intravenous (i.v.) administration of the respective drug, in daily repeated experiments in individual rats. The calculated area-under-curve (AUC) values reveal an effect of the tonicity and temperature on the concentrations of the drugs measured in the dialysate.

Abbreviations

AUC	=	area under the concentration-time curve
PSI	=	post-surgery intervals
ECF	=	extracellular fluid
$C_{brain\ ECF}$	=	concentration of drugs in ECF
C_{dial}	=	concentration of drug in dialysate
iso	=	isotonic perfusate
hypo	=	hypotonic perfusate

INTRODUCTION

Intracerebral microdialysis has been developed as a method to obtain information on the composition of brain extracellular fluid. In many ways it is still a method in development, and it should be realized that different experimental conditions used in microdialysis experiments may lead to different outcomes (Benveniste and Diemer, 1987; Benveniste *et al.*, 1987, 1989). The following factors are crucial in this respect: probe geometry (Westerink and De Vries, 1988), membrane material (Tao and Hjort, 1992), composition of the perfusing solution (Moghaddam and Bunney, 1989), flow rate/dialysis surface (Tossman and Ungerstedt, 1986; Lindefors *et al.*, 1989), surgical trauma/periprobe tissue reactions (Yergey *et al.*, 1990), anaesthesia (Hamilton *et al.*, 1992), repeated experiments (Westerink and Tuinte, 1986), and the rates of processes such as distribution, metabolism, and elimation that determine the concentration of the compound of interest in the tissue under investigation (Morrison *et al.*, 1992). Most of the investigations on the influence of experimental conditions on the microdialysis data obtained have been performed in the neurochemical field, and it is clear that for other applications these factors should also to be taken into account.

Our interest is to use this technique to measure drug concentrations in normal and diseased brain. In our initial experimental setup, a post-surgery interval (PSI) of 24 h was chosen, because the period between 24–48 h post surgery was found to be an optimal experimental condition; the brain tissue would have recovered from the initial reactions of the implantation of the microdialysis probe as measured by the formation of eicanosoids (Yergey and Heyes, 1990) and changes in cerebral blood flow (Benveniste *et al.*, 1987), while long-term reactions may not be of importance at this stage (Benveniste *et al.*, 1987; Tossman and Ungerstedt, 1986; Westerink and Tuinte, 1988). We have chosen as a perfusion medium the phosphate-buffered solution prepared in accordance with the method described by Moghaddam and Bunney (1989). They have shown that the solution that was initially used in microdialysis experiments (Ringer's) contained a too high concentration of Ca^{2+} (3.4 mM) as compared to the Ca^{2+} content in brain ECF (1.2 mM). Such a difference would affect the basal as well as the pharmacological stimulated release of dopamine.

However, the effect of perfusate temperature had not been addressed in the above-mentioned reports. Therefore, in this study the influence of the perfusate temperature (just before entering the microdialysis probe) was investigated. The effect of a hypotonic perfusate, expected to induce a kind of diseased state, was also examined. Because we were interested in longitudinal experiments, the possibility of repeated experiments in the same animal was checked with the isotonic as well as with the hypotonic perfusate. In these studies the AUC values of either acetaminophen and atenolol at the different experimental conditions were used for data comparisons.

MATERIALS AND METHODS

Animals

Adult male SPF Wistar rats (body weight 180–250 g) from the Sylvius laboratories, University of Leiden, were used. The rats were housed in Macrolon™ cages with standard hardwood bedding and free access to water and standard laboratory rat diet (RMH-TH, Hope Farms, Woerden, The Netherlands).

Surgery

For microdialysis the rats were anaesthetized with an intramuscular injection of 150 μl of Hypnorm® (Janssen Pharmaceutica, Goirle, The Netherlands) and placed in a stereotaxic frame. Incisions were made to expose the skull which was locally anaesthetized thereafter with a 0.6% solution of lidocaine. Holes (1.5 mm) were drilled in the lateral plane of the skull, allowing the horizontal introduction of a dialysis probe, using a tungsten wire (TW5-3, Clark Electro Medical Instruments, England), through the cortex at 2 mm below bregma. The dialysis fibre (O.D. 0.29 mm, C-DAK artificial kidney 201-800 D 135 SCE, CD Medical B.V., Rotterdam, The Netherlands) was covered with silicone glue (Rhodosil CAF 3, Rhône-Poulenc, Amstelveen, The Netherlands) except for a 10 mm central zone (or a smaller length, depending on the study design). Stainless steel needles (Microlance 25G inlet, 23G outlet, Becton Dickinson B.V., The Netherlands), glued to both ends of the dialysis fibre with pattex gel (Henkel Nederland B.V., Nieuwegein, The Netherlands), were secured with dental cement on the top of the skull. The animals were allowed to recover from probe implantation and anaesthesia in the experimental room, for a period of 24 h before the start of the first experiment.

A perfusate temperature of 38 °C was achieved by the use of a subcutaneous cannula (polyethylene tube, I.D. 0.58 mm, length of about 20 cm) at the back of the rat. The perfusate fluid was led through this cannula just before entering the microdialysis probe in the brain, allowing the fluid to equilibrate to rat body temperature.

For i.v. drug administration and serial blood sampling, polyethylene cannulas (I.D. 0.58 mm, O.D. 0.8 mm) filled with 20 E of heparin were implanted into the femoral vein and femoral artery respectively under ether anaesthesia. The recovery period was at least 2 h before the start of the experiment.

Experimental Procedure

For transcortical microdialysis experiments, the stainless steel needles at both sides of the microdialysis probe were connected by means of polyethylene tubing (O.D. 0.51 mm, I.D. 0.08 mm, 80 cm) to a perfusion pump (Gilson) and a sample loop respectively. The cage of the rat was put at a higher position than the outlet of the loop in order to prevent additional fluid pressure in the dialysis membrane. The dialysis probe was perfused at 7 μl/min with a helium-degassed 2 mM phosphate buffer containing 145 mM sodium, 2.7 mM potassium, 1.2 mM calcium, 150 mM chloride, 1.0 mM magnesium and 0.2 mM ascorbate, pH = 7.4. The hypotonic perfusion solution used was identical to the above mentioned isotonic perfusate, except for the sodium and chloride concentrations which were 14.5 and 15 mM respectively.

The rats were dialysed with the buffer solution for 30 min to obtain dialysis equilibrium and blank data. Subsequently the drug was administered and dialysate samples were measured for 120 min. During the course of the experiment the rats were freely moving and had free access to water and food.

For i.v. administration, the drug solution was injected into the femoral vein over 1 min. A solution of 825 μg (5.4 μmol) of acetaminophen or 10 mg (38 μmol) of atenolol in 500 μl saline was administered. At regular intervals (0, 5, 15, 30, 45, 60, 90, and 120 min) blood samples of 200 μl were drawn from the femoral artery and collected in heparinized tubes. Plasma was obtained by centrifugation and stored at −20 °C until analysis.

Experiments were performed to investigate the intra-animal reproducibility over days: in consecutive experiments individual rats were used repeatedly every day, with the experiments starting at post-surgery intervals of 24, 48 and 72 h.

Drug Analysis

Acetaminophen

The acetaminophen HPLC system consisted of a reversed phase column (Spherisorb, 10 cm * 4.6 mm I.D., S3 ODS 2, Phase Separations, Waddinxveen, The Netherlands), a precolumn (Pellicular reverse phase, Chrompack) and a electrochemical detector (glassy carbon electrode with an oxidation potential of 800 mV versus an Ag/AgCl electrode, Antec Leyden B.V., The Netherlands). The mobile phase consisted of a mixture of a 20 mM sodium phosphate and a 50 mM sodium citrate buffer containing 0.25 mM sodium octane sulphonic acid and 15% (v/v) methanol, pH = 3.0. The flow was 1.0 ml/min. Chromatographic data were recorded and processed with a SP4100 computing integrator (Spectra Physics, SP4100 computing injector, Spectra Physics B.V., Eindhoven, The Netherlands). The interassay variability in aqueous solution was less than 3% ($n = 15$), with a detection limit of 0.1 ng/ml (66 fmol). All assay calibration curves showed good linearity (correlation coefficient $r > 0.9995$). The microdialysate was on-line injected, up to 120 min after the administration of the drug, into the HPLC system through a Valco injection valve equipped with a 10 μl loop, with a repetition time of 3.5–4.0 min.

Atenolol

The atenolol HPLC system consisted of a reversed phase column (Spherisorb, 10 cm * 4.6 mm I.D., S3 ODS 2, Phase Separations, Waddinxveen, The Netherlands), a precolumn (Pellicular reverse phase, Chrompack, Middelburg, The Netherlands) and a Shimadzu RF-350 fluorimeter (Shimadzu Corp., Kyoto, Japan) with an excitation wavelength of 276 nm and an emission wavelength of 309 nm. The mobile phase consisted of 73.5% (v/v) sodium acetate buffer, pH = 4.0, containing 5 mM of sodium octane sulphonate and 26.5% (v/v) of acetonitrile. The flow was 1.0 ml/min. The inter-assay variation in aqueous solution was less than 7% ($n = 15$), with a detection limit of 100 ng/ml (6 pmol). All assay calibration curves showed good linearity ($r > 0.9990$). The microdialysate was on-line injected, up to 120 min after the administration of the drug, into the HPLC system by a Valco injection valve equipped with a 16 μl loop, with a repetition time of 3.5–4.0 min.

Data Analysis

The profiles obtained from the dialysate were corrected for the time needed to transport the dialysate into the HPLC system (lag-time). To estimate $C_{brain\ ECF}$,

TABLE 1 Effect of perfusate temperature on AUC of acetaminophen and atenolol after i.v. administration (iso = isotonic, hypo = hypotonic perfusate, post-surgery interval of 24 h). Values are presented as mean ± SEM. * = significantly different (Student's *t*-test, $P < 0.05$) vs. corresponding values at 38 °C.

Acetaminophen		iso		hypo
Perfusion temperature °C	38	24	38	24
Atenolol AUC ($\mu g/min/ml$)	65.5±6.7	62.3±11.3	n.d.	n.d.
Acetaminophen AUC ($\mu g/min/ml$)	16.4±1.9	16.9±1.3	4.1±0.81	7.8±0.85*

the C_{dial} of the drugs were corrected for *in vitro* recovery at 37 °C for a 10 mm dialysis zone in a non-stirred solution of the drug in the perfusate buffer. This was (mean ± SEM) 23 ± 1% for acetaminophen and 13 ± 1% for atenolol ($n = 3$ probes, 3 determinations per probe).

Individual $C_{brain\ ECF}$ of acetaminophen and atenolol was used to calculate AUC values (0–120 min) by means of the trapezoidal rule (Siphar, modeling package, SIMED, Creteil, France). These AUC values were used for comparative analysis of the results. Statistical evaluation was performed with Student's *t*-test (to compare mean values) or Duncan's Multiple rank test (for paired data in the repeated experiments). The value of $P < 0.05$ was considered statistically significant.

RESULTS

Table 1 shows that the temperature of the perfusate does not affect the AUC values obtained with the isotonic dialysate. However, when the hypotonic solution was used, acetaminophen AUC values were two-fold higher for the perfusate administered at room temperature (24 °C).

Table 2 shows the influence of the tonicity of the dialysis solution on the AUC values when the studies were conducted according to a longitudinal design with repeated experiments on three consecutive days. It shows that considerable differences in AUC values can be obtained if an inappropriate perfusion solution is used, and this pattern is most pronounced for the drug atenolol. It should be mentioned, in addition, that there was no difference in the values obtained from rats in which the implantation had taken place in the morning, afternoon or evening.

DISCUSSION

The present study shows that the temperature and tonicity of the perfusate are two important factors in the dialysis process. Although the temperature effect is not apparent under 'normal' conditions, it is clear that under 'stressed' tissue conditions

TABLE 2 Effect of perfusate composition on AUC values of acetaminophen and atenolol after i.v. administration (iso = isotonic, hypo = hypotonic perfusate of 38 °C). Values are presented as mean ± SEM ($n = 6$). * = significantly different (Student's t-test, $P < 0.05$) compared to the accompanying AUC values for a post-surgery interval of 24 h.

	iso			hypo		
Acetaminophen						
Post-surgery interval (h)	24	48	72	24	48	72
AUC (μg/min/ml)	13.8±0.52	13.4±0.74	10.5±0.64*	7.94±0.70	12.3±1.69*	9.08±1.04
Atenolol						
Post-surgery interval (h)	24	48	72	24	48	72
AUC (μg/min/ml)	62.4±12.9	64.0±10.1	61.3±11.3	54.4±13.6	206±13.3*	232±12.9*

the periprobe tissue is not able to counterbalance (fully) the resulting temperature difference (between the 24 and 38 °C perfusate) at the level of the dialysing part of the probe. Because the tissue under investigation supposedly is 'stressed' especially in the pathological state , the use of the prewarmed perfusate buffer is recommended in such investigations.

The use of the hypotonic perfusate, which would obviously not be chosen if not to investigate the flexibility of the tissue studied, had a dramatic effect on the AUC values obtained in three runs of the experiments in an individual animal. Therefore, the situation in which the use of the isotonic perfusate lead to more or less reproducible AUC values on the consecutive days, strongly indicated that in principle this technique can be used in subchronic experiments. However, for each drug to be used in longitudinal studies, the time span of reproducibility of the experimental parameters should be checked.

CONCLUDING REMARKS

An intracerebral microdialysis technique is used in freely-moving rats to measure the concentration of atenolol and acetaminophen in brain dialysate. The same procedure can also be used to measure the concentration of drugs such as methotrexate and dibutylmethotrexate, which are used in the treatment of leukemia and other forms of cancer. Indeed, it is our interest to use this technique to measure drug concentrations in normal and diseased brains.

REFERENCES

Benveniste, H. and Diemer, N.H. (1987) Cellular reactions to implantation of a microdialysis tube in the rat hippocampus. *Acta Neuropath. (Berl.)*, **74**, 234–238.
Benveniste, H., Drejer, A., Shoesboe, A. and Diemer, N.H. (1987) Regional cerebral glucose phosphory-lation and blood flow after insertion of a microdialysis fiber through the dorsal hippocampus in the rat. *J. Neurochem.*, **49**, 729–734.

Benveniste, H., Hansen, A.J. and Ottosen, N.S. (1989) Determination of brain interstitial concentrations by microdialysis. *J. Neurochem.*, **52**, 1741–1750.

Hamilton, M.E., Mele, A. and Pert, A. (1992) Striatal extracellular dopamine in conscious vs. anaesthetized rats: Effects of chloral hydrate anaesthetic on responses to drugs of different classes. *Brain Res.*, **597**, 1–7.

Moghaddam, B. and Bunney, B.S. (1989) Ionic composition of microdialysis perfusing solution alters the pharmacological responsiveness and basal outflow of striatal dopamine. *J. Neurochem.*, **53**, 652–654.

Morrison, P.F., Bungay, P.M., Hsiao, J.K., Mefford, I.V., Dijkstra, K.H. and Dedrick, R.L. (1992) Quantitative microdialysis. In *Microdialysis in Neuroscience*, Robinson, T.E. and Justice, J.B. (Eds.), pp. 49–79.

Lindefors, N., Armberg, G. and Ungerstedt, U. (1989) Intracerebral microdialysis: I. Experimental studies of diffusion kinetics. II. Mathematical studies of diffusion kinetics. *J. Pharmacol. Meth.*, **22**, 141–183.

Tao, R. and Hjort, S. (1992) Differences in the *in vitro* and *in vivo* 5-HT extraction performance among three common microdialysis membranes. *J. Neurochem.*, **59**, 1778–1785.

Tossman, U. and Ungerstedt, U. (1986) Microdialysis in the study of extracellular levels of amino acids in the rat brain. *Acta Physiol. Scand.*, **128**, 9–14.

Westerink, B.H.C. and De Vries, J.B. (1988) Characterization of *in vivo* dopamine release as determined by brain microdialysis after acute and subchronic implantations: methodological aspects. *J. Neurosci.*, **51**, 683–687.

Westerink, B.H.C. and Tuinte, M.H.J. (1986) Chronic use of intracerebral microdialysis for the *in vivo* measurements of 3,4-dihydroxyphenylacetic acid. *J. Neurochem.*, **46**, 181–189.

Yergey, J.A. and Heyes, M.P. (1990) Brain eicosanoid formation following acute penetration injury as studied by *in vivo* microdialysis. *J. Cerebr. Blood Flow Metab.*, **10**, 143–146.

III.1 BRAIN MICRODIALYSIS IN THE MOUSE: DETERMINATION OF BIOGENIC AMINE METABOLITES IN THE DORSAL HIPPOCAMPUS AND THE NUCLEUS ACCUMBENS*

N. LAUNAY, G. BOSCHI,[1] R. RIPS AND J.-M. SCHERRMANN

INSERM U 26, Hôpital Fernand Widal, 200 rue du Faubourg Saint-Denis, 75475 Paris Cedex 10, France

Microdialysis of small brain areas of OF1 mice is shown to be feasible using the smallest commercially available probes (CMA/11). The brain areas studied were the dorsal hippocampus and nucleus accumbens. The basal concentrations of catecholamines, indolamines and their metabolites in dialysate samples were measured by HPLC using electrochemical detection. This procedure allows a simultaneous correlation of the neurobiochemical changes and pharmacological responses, and facilitates further biochemical and pharmacokinetic research in the mouse.

INTRODUCTION

The technique of microdialysis has been most widely used in the rat. However, there are frequently species differences in the pharmacological and biochemical effects of drugs. Microdialysis was recently used to investigate the distribution of drugs in the brains of rabbits (Wang *et al.*, 1993). There have only been two studies using microdialysis in mice; both showed species differences in the metabolism of catecholamines and indolamines in the striatum of rats and mice (Wood *et al.*, 1988; Rollema *et al.*, 1989). However, the striatum of the mouse is large enough for dialysis probes to be implanted and used for the collection of the dialysates. Further studies, however, require microdialysis in other brain areas, such as the hippocampus and the nucleus accumbens. The present work was undertaken to adapt the microdialysis technique to these small cerebral regions of the mouse, in order to carry out pharmacokinetic and biochemical studies in this species. The feasibility of this methodology in mice was assessed by measuring concentrations of biogenic amines and their metabolites in dialysates from the dorsal hippocampus and nucleus accumbens by HPLC using electrochemical detection.

*Part of this work was presented at the Third International Symposium on Microdialysis and Allied Analytical Techniques, May 1993, Indianapolis, USA.

[1] Corresponding author.

MATERIALS AND METHODS

Animals and Chemicals

Animals

All studies were carried out using male OF1 mice (Iffa Credo) of 7 weeks old and weighing 30–32 g on the day of the experiment. After surgery, they were housed individually and given water and food *ad libitum*. Each mouse was used only once.

Guide cannulae and microdialysis probes

Guide cannulae were prepared from hypodermic needles. CMA/11 probes (outer diameter 0.24 mm, length 1 mm) with a polycarbonate-polyether copolymeric membrane and a molecular cut-off of 20,000 Daltons were used (Carnegie Medicin AB, Stockholm, Sweden). Each probe was used several times if there was no variation in the *in vitro* recoveries of the analytes.

Recovery in vitro

Probes were placed in a 100 pmol/ml (10^{-7} M) solution of the mixed standards diluted in Ringer's solution and perfused with normal Ringer's solution for 30 min. The dialysates were collected under the same conditions as during *in vivo* micro-dialysis.

Compounds

Noradrenaline (NA), 3-methoxy, 4-hydroxyphenylethyleneglycol (MHPG), dopamine (DA), dihydroxyphenylacetic acid (DOPAC), homovanillic acid (HVA), 3-methoxy tyramine (3-MT), 5-hydroxyindoleacetic acid (5-HIAA), 5-hydroxytryptamine (5-HT) were purchased from Sigma. Stock solutions (10^{-3} M) of the standards were prepared in 0.1 M perchloric acid and kept at 4 °C. Diluted standards (0.2×10^{-7} M) in Ringer's solution were prepared daily.

Experimental Procedures

Surgical procedures

Animals were anaesthetized with 5% chloral hydrate (500 mg/kg i.p.) and their skulls were shaved. They were placed in a stereotaxic apparatus, and the stereotaxic coordinates for the guide cannula were determined from the atlas of the rat brain (König and Klippel, 1963) for general orientation. The target point was fixed for the dorsal hippocampus (in mm): 1.9 anterior to lambda, 1.4 lateral to sagittal suture,

–2.1 ventral to skull surface; for the nucleus accumbens (in mm): 1.3 anterior to bregma, 1.1 lateral to sagittal suture, –4.4 ventral to skull surface. The guide cannula was inserted into the cortex (to minimize the tissue damage) and fixed with dental cement, as previously described (Boschi et al., 1981). The mice were allowed to recover from surgery for 4–5 days.

Microdialysis procedure

Microdialysis was performed between 1 pm and 6 pm. On the day of the experiment, mice were placed in a bowl cage with a collar attached to a holder connected to a swivel (CMA/120 system). The dialysate collection was started immediately and continued for 5 h. The first sample covered the 20 min immediately after the probe insertion. The second sample was collected for the first 20 min of the second hour. Subsequent samples were collected every 20 min for 3 hr to determine the baseline and to check its stability.

The inflow to the microdialysis probe was driven by a CMA/100 microinjection pump, and the outflow was collected in small polypropylene tubes; both were connected by FEP tubing. The perfusion fluid (Ringer's solution) contained (in mM): NaCl, 147; KCl, 4; $CaCl_2$, 2.4; pH 6.4. The flow rate was 1.5 μl/min. A 25 μl aliquot of the sample was injected immediately into the HPLC system and analysed for catecholamines, indolamines and their metabolites.

HPLC analysis

The system consisted of a Hitachi 6200 pump (Merck), a Rheodyne valve (50 μl loop), a C8 reverse-phase column (Lichrospher 60 RP Select B, Merck), 125 × 4 mm, 5 μm, and an LC 17 electrochemical detector with a glassy carbon electrode (BAS). The working electrode was set at +0.8 V with respect to an Ag/AgCl reference electrode, range 0.2 nA. The mobile phase contained 0.1 M KH_2PO_4, 0.1 mM EDTA, 5 mM Pic B7 (heptane sulphonic acid) and 9% methanol, pH 4.1. The flow rate was maintained at 0.7 ml/min for hippocampal dialysates and at 0.8 ml/min for the dialysates from the nucleus accumbens. The limit of detection (S/N = 3) was about 20 fmol. All solutions (Ringer's solution, KH_2PO_4 buffer) were filtered through a 0.2 μm membrane filter. The amount of each compound in dialysate was determined by comparison with the peak areas of standards run with each experiment.

Histology

The mouse was killed at the end of the experiment, the brain was removed and placed in 30% commercial formalin. Frontal sections (120 μm) were cut and examined to check the location of the dialysis site (by comparison with sections of the rat brain illustrated in the atlas of König and Klippel (1963). Only data from animals with correct dialysis probe placement were used.

TABLE 1 Relative recovery (*RR*) of CMA/11 probes for neurotransmitters and their metabolites in a standard mixture containing 10^{-7} M of each analyte in Ringer's solution.

Fraction	NA	MHPG	DOPAC	DA	5-HIAA	HVA	5-HT
1	13.2±3.7	22.8±0.6	11.0±1.8	9.8±1.9	7.9±1.4	12.8±2.5	6.8±0.6
	(5)	(2)	(7)	(6)	(7)	(6)	(7)
2	13.5±2.8	26.1±2.3	9.2±1.3	9.3±2.0	7.6±1.5	12.8±2.2	6.6±0.6
	(5)	(2)	(7)	(6)	(7)	(6)	(7)
3	14.0±2.9	18.4±1.8	7.9±1.0	8.9±2.0	7.4±1.4	11.0±1.7	7.1±0.8
	(5)	(2)	(6)	(5)	(6)	(5)	(6)
4	12.0±2.7	16.3±2.9	6.5±0.7	8.6±2.6	7.1±1.4	11.3±2.9	6.2±0.8
	(5)	(2)	(6)	(5)	(6)	(5)	(6)
5	10.2±2.3	15.7±2.9	7.8±1.4	8.6±1.8	8.6±1.6	13.3±2.5	7.2±0.8
	(4)	(2)	(6)	(5)	(6)	(5)	(6)
6	10.0±2.3	15.9±3.8	8.3±1.6	8.0±1.6	8.0±1.6	14.6±3.4	7.9±0.9
	(4)	(2)	(5)	(6)	(6)	(5)	(6)

CMA/11 microdialysis probes (0.24 mm O.D., 1 mm length) were perfused with Ringer's solution at 1.5 μl/min (consecutive fractions were collected at 20 min intervals). *RR* (%) is expressed as the mean ± SEM (number of probes).

Statistical analysis

Results are expressed as means ± SEM. They were analysed using a multivariate analysis of variance (ANOVA) with repeated measures and followed, when appropriate, by Bonferroni's test.

RESULTS

In Vitro Neurotransmitter Extraction by CMA/11 Probes

The CMA/11 probes (0.24 mm O.D., 1 mm length) are the smallest commercially available probes. They are very fragile because they tend to dry out and leak. They must therefore be handled with great care. A new microdialysis probe should be tested before use. The performance of a probe was estimated by measuring its relative recovery (*RR*) *in vitro*. This is the ratio of the concentration of a compound in a dialysate fraction (*Cd*) over the concentration of this compound in the sample solution (*Cs*) i.e. *RR= Cd/Cs*. *RR* determinations were started 30 min after immersion of the probe. As shown in Table 1, the CMA/11 probes displayed a stable relative recovery for each analyte throughout the test, except for MHPG, for which the RR gradually declined.

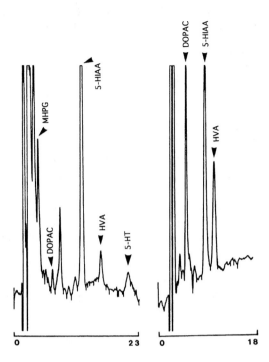

Figure 1 Chromatograms of 25 µl of dialysate from the (left) dorsal hippocampus: MHPG, DOPAC, 5-HIAA, HVA, 5-HT (697, 38, 908, 270, 48 fmol, respectively) and from the (right) nucleus accumbens: DOPAC, 5-HIAA, HVA (1763, 1538, 2107 fmol, respectively) at the steady-state. Abscissa: time in min. For experimental conditions, see Materials and Methods.

Basal Concentrations of Catecholamines, Indolamines and their Metabolites in Dialysates

Figure 1 shows chromatograms of dialysates from the dorsal hippocampus and nucleus accumbens. The identity of the MHPG peak was confirmed by adding a small amount of the standard to brain dialysate samples.

Figure 2 shows the concentrations (pmol/30 µl) of biogenic amines and their metabolites in dialysates collected from the dorsal hippocampus of conscious mice. The outputs of MHPG, DOPAC and HVA were constant over the time course of the experiment. In contrast, the concentration of 5-HIAA progressively declined from 1.10 ± 0.10 pmol/30 µl to 0.68 ± 0.05 pmol/30 µl ($p < 0.05$) 3 h later without ever reaching a constant level. The first dialysate sample from the dorsal hippocampus, collected immediately after the probe insertion, detected a substantial amount of 5-HT (0.39 ± 0.07, $p < 0.002$), probably due to tissue damage during the introduction of the probe. The output of 5-HT rapidly reached its basal level and 3 h later was close to the limit of detection, before dropping to zero. DA remained undetectable.

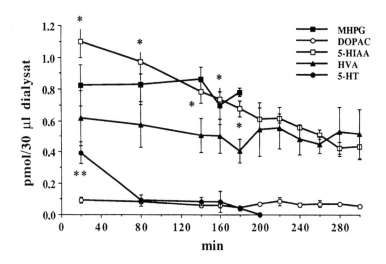

Figure 2 Concentrations of MHPG, DOPAC, 5-HIAA, HVA and 5-HT in dialysates collected from the dorsal hippocampus of conscious mice over 3–5 h after the insertion of the probe. Each point is the mean ± SEM of a number of values (compounds were undetectable towards the end of the experiment in some mice). Multivariate analysis of variance (ANOVA) with repeated measures followed by Bonferroni's test. *$p < 0.05$, **$p < 0.002$. For each curve, the first point was compared to the second one, the second point to the third one and so on.

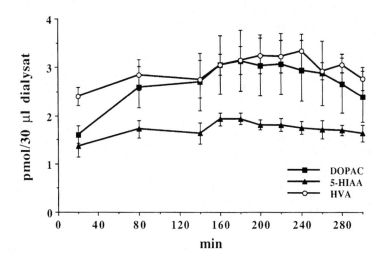

Figure 3 Concentrations of DOPAC, 5-HIAA and HVA in dialysates collected from the nucleus accumbens over 5 h following probe insertion. Each point is the mean ± SEM for 5 mice. Multivariate analysis of variance (ANOVA) with repeated measures (see Legend to Figure 2).

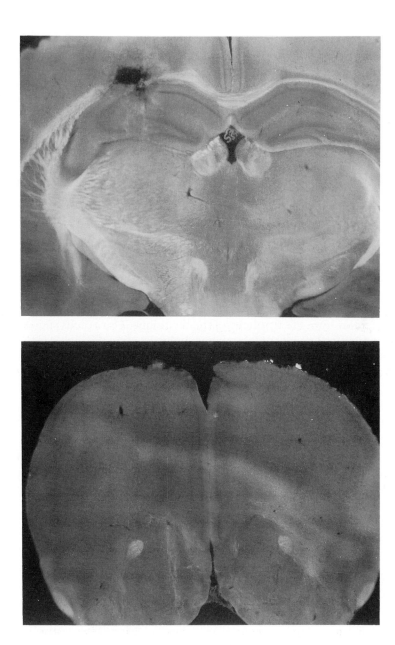

Figure 4 Photomicrographs of a coronal section (120 μm) through the dorsal hippocampus (upper) and the nucleus accumbens (bottom) of the mouse showing the implantation site of the dialysis probe.

The DOPAC and HVA levels in dialysates from the nucleus accumbens stabilized within 80 min and remained constant until the end of the experiment (Figure 3). The concentration of 5-HIAA was stable from the beginning of dialysis. DA was scarcely detectable (data not shown). 3-MT remained undetectable.

Figure 4 shows the dialysis probe placement in the dorsal hippocampus and the nucleus accumbens.

DISCUSSION

These experiments show that microdialysis can be performed in small brain areas of unrestrained (freely moving), conscious mice. This technology requires the use of very small, delicate dialysis probes. We used the smallest commercially available CMA probes for rat brain microdialysis (CMA/11) and thus minimizes methodological differences due to the probe construction and therefore makes this technology more accessible.

Microdialysis was successfully performed in two areas: the dorsal hippocampus and the smaller nucleus accumbens. The measure of the concentrations of neurotransmitters and their metabolites in these regions by HPLC with electrochemical detection demonstrate the feasibility of the technique. Our results show no significant fluctuations in the biogenic amine metabolite levels in the dialysates, except for the level of 5-HIAA in the hippocampal dialysates. This means that the metabolite concentrations in the dialysates represent stable extracellular concentrations.

CONCLUDING REMARKS

Brain microdialysis in the mouse could contribute greatly to pharmacokinetic and biochemical research in this species. This may be especially important in transgenic mice, or in mdr1a-deficient mice.

REFERENCES

Boschi, G., Launay, N. and Rips, R. (1981) Implantation of an intracerebral cannula in the mouse. *J. Pharmacol. Meth.*, **6**, 193–198.

König, J.E.R. and Klippel, R.A. (1963) *The Rat Brain: A Stereotaxic Atlas of the Forebrain and Lower Parts of the Brainstem.* Williams and Wilkins, Baltimore.

Rollema, H., Alexander, G.M., Grothusen, J.R., Matos, F.F. and Castagnoli, Jr.N. (1989) Comparison of the effects of intracerebrally administered MPP+ (1-methyl-4-phenylpyridinium) in three species: Microdialysis of dopamine and metabolites in mouse, rat and monkey striatum. *Neurosci. Lett.*, **106**, 275–281.

Wang, Y., Wong, S. and Sawchuk, R.J. (1993) Microdialysis calibration using retrodialysis and zero-net flux: Application to a study of the distribution of ziduvidine to rabbit cerebrospinal fluid and thalamus. *Pharm. Res.*, **10**, 1411–1419.

Wood, P.L., Kim, H.S., Stocklin, K. and Rao, T.S. (1988) Dynamics of the striatal 3-MT pool in rat and mouse: Species differences as assessed by steady-state measurements and intracerebral dialysis. *Life Sci.*, **42**, 2275–2281.

IV.1 MICRODIALYSIS OF SUBCORTICAL STRUCTURES IN CONSCIOUS CHRONIC RABBITS

W.Z. TRACZYK, M. ORLOWSKA-MAJDAK, A. WALCZEWSKA
AND B. DZIEDZIC

*Department of Physiology, Institute of Physiology and Biochemistry,
Medical University of Lodz, ul. Lindleya 3, 90-131 Lodz, Poland*

The technique of implantation of stereotaxically oriented microdialysis probes into subcortical structures, that is into hypothalamus, hippocampus and caudate nucleus in rabbits is described. The potential use of microdialysis probes CMA/Microdialysis AB permanently implanted into subcortical structures was maintained in some animals for up to three months.

INTRODUCTION

The microdialysis technique allows us to study the concentration of neurotransmitters and neuromodulators in the extracellular fluid of particular brain structures. Each brain structure has a specific composition of cell bodies and nerve endings operating with different transmitters and modulators. This specificity of brain structures affects the composition and concentration of active compounds of extracellular fluid in particular brain structures.

The microdialysis technique has some advantages in comparison, for example, to microperfusion of the brain structures by push-pull techniques. A continuous microdialysate flow is independent of the changes in intracranial pressure. The two techniques still have a common disadvantage — damage of some nerve cell bodies and nerve cell processes by the insertion of microdialysis probes. Reactive gliosis appeared shortly after brain damage and the maximum response was seen during the first to third week post-lesion (Kitamura *et al.*, 1978; Morshead and van der Kooy, 1990). It is obvious that the membrane of microdialysis probes may be surrounded by a proliferative process which can significantly modify the composition of extracellular fluid and affect the concentration of physiologically active compounds in the outflowing microdialysis fluid.

The disadvantage could probably be overcome by the insertion of the microdialysis probe only once into a subcortical structure and by waiting until the proliferative response to subside after several weeks. For this purpose an experimental technique was developed which allows the use of the implanted microdialysis probes for several months.

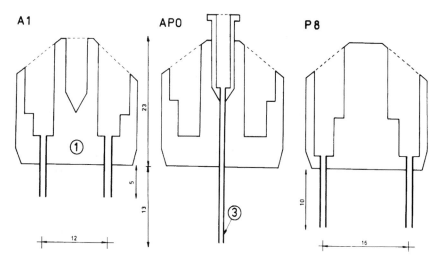

Figure 1 Frontal cross-sections of a plexiglass headpiece (1) with vertical guide cannulae. The APO cross-section shows the guide cannula (3) to the 3rd cerebral ventricle. The A1 and P8 cross-sections show the guide cannulae for implantation of the microdialysis probes: the A1 into the caudate nuclei and P8 into the hippocampus on both sides. All dimensions are in millimeters on all figures.

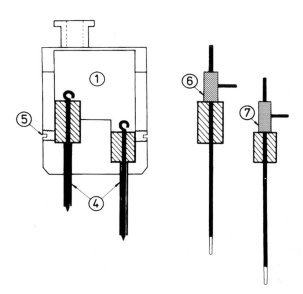

Figure 2 The sagittal cross-section of plexiglass headpiece (1) with stilettes in the pistons introduced into the guide cannulae (4). With screws pressing on the pistons the stilettes are fixed in the guide cannulae.

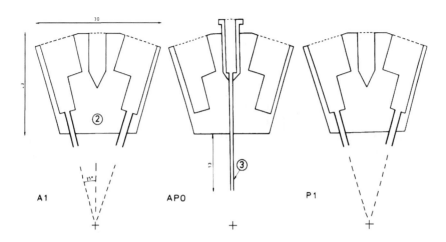

Figure 3 Frontal cross-sections of plexiglass headpiece (2) with guide cannulae at 15 ° angle to the vertical 3rd cerebral ventricle guide cannula (3). The guide cannulae are oriented at an angle to reach the hypothalamus with the tip of the microdialysis probes on both sides in two frontal planes A1 and P1.

MATERIALS AND METHODS

Animals and Surgery

Adult male or female Chinchilla rabbits were housed in single cages in a temperature-controlled and light-regulated (lights on 06.00 h and off 20.00 h) room for at least 30 days prior to any surgical intervention. Female rabbits underwent bilateral ovariectomy as the initial surgical intervention. Some weeks later female ovariectomized rabbits and the male rabbits were subjected to cranial surgery. Deep surgical anaesthesia was induced by subcutaneous injections of atropine sulphate (1.0 mg per animal) and morphine hydrochloride (7.5 mg per kg b.w.) followed by i.v. slow injection of hexobarbital-natrium (Germed; 40.0 mg per kg b.w.). After reaching deep surgical anaesthesia each animal was mounted on the stereotaxic frame (Sawyer *et al.*, 1954) and stereotaxically implanted with a plexiglass headpiece with guide cannulae. Male rabbits were implanted with headpieces with vertical guide cannulae, one for 3rd cerebral ventricle and two for caudate nuclei and two others for hippocampi (Figures 1 and 2).

Female rabbits were implanted with headpieces with one vertical guide cannula for 3rd cerebral ventricle and four others at a 15 ° angle to the vertical cannula, leading to the hypothalamus (Figure 3). Push-pull cannulae were implanted into the hypothalamus with the same angle in dogs (Traczyk, 1966) and rabbits (Traczyk *et al.*, 1992).

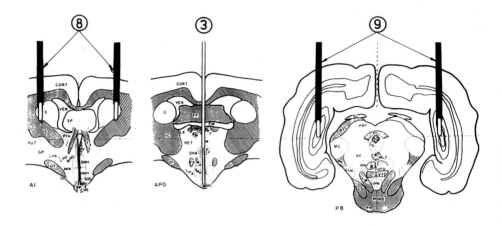

Figure 4 Frontal section of rabbit brain A1, APO and P8 with the position of microdialysis probes in the caudate nuclei (8) or hippocampi (9) and 3rd ventricle cannula (3). (Modified frontal plane from Sawyer *et al.* atlas with author's permission).

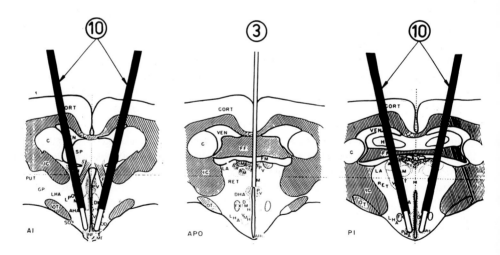

Figure 5 Frontal sections of rabbit brain A1, APO and P1 with position of microdialysis probes in the hypothalamus (10) and 3rd cerebral ventricle guide cannula (3). (Modified frontal plane from Sawyer *et al.* atlas with author's permission).

Five holes were made using a dental drill in the skull bones in places corresponding to the rabbit stereotaxic atlas (Sawyer *et al.*, 1954). Four small holes were drilled for the guide cannulae leading to the caudate nuclei and hippocampi in male rabbits or to the hypothalamus in female rabbits (Figures 4 and 5). The fifth hole in the skull bones around the Bregma was enlarged to approximately 4×4 mm, to expose dura mater and the superior sagittal sinus. After cutting the dura under operation microscope, the tip of the 3rd ventricle guide cannula should be lowered stereotaxically exactly in the middle between the cerebral hemispheres. The dura and sinus were gently drawn to one side to expose the space between the hemispheres. The tip of the 3rd ventricle cannula with the whole headpiece was lowered stereotaxically 10 mm deep from the surface of the dura mater, in the APO frontal plane (Figures 4 and 5). The tip of the 3rd cerebral ventricle guide cannula was used as a reference point.

After introducing the guide cannula into the 3rd ventricle the closure cap with stilette was removed and the cerebrospinal fluid meniscus movements were observed under the operation microscope. Up and down respiratory movements of the cerebrospinal fluid meniscus in the 3rd ventricle guide cannula confirmed the correct position of the tip of the 3rd ventricle guide cannula and at the same time all other guide cannulae.

Four stainless steel screws were screwed into the skull bones, two at the front and the other two at the rear of the headpiece. The screws and the lower part of the headpiece were covered with dental cement (Duracril, Spofa). After hardening of the dental cement a few sutures were put on the skin on both sides of the headpiece, the stand for coil and aluminium cover were fixed to the headpiece (Figures 6 and 7). After the surgery each rabbit received intramuscular injections of 100,000 IU of benzylpenicillin potassium (Polfa-Tarchomin) and 0.5 g of streptomycin (Polfa-Tarchomin) daily during five consecutive days.

The length of the guide cannulae for subcortical structures filled with a stilette was chosen so that when the tip of the guide cannula reached stereotaxically the 3rd ventricle, at the same time all other guide cannulae with stilettes inside should puncture the dura mater.

With headpiece implanted into the skull and guide cannulae fitted with the stilettes, the animals can be kept in the animal house for several months. They should then be accustomed to spend several hours daily in a special box designed to facilitate microdialysis. The box allowed free access to food and water but restricted body rotation.

Microdialysis

Before introducing the microdialysis probe into subcortical structures, the rabbits were put into experimental boxes and 20% solution of mannitol was i.v. infused at the amount of 600 mg per kg b.w. of the animal during 1 h. Then the screws fixing the pistons with stilettes inside were removed from the guide cannulae. Under the operation microscope the microdialysis probes were inserted (Figure 8) into the

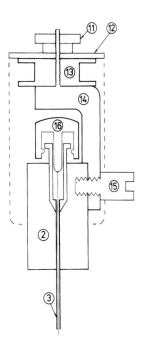

Figure 6 Sagittal cross-section of plexiglass headpiece (2) with coil for inlet and outlet tubings (13) and cover (12).

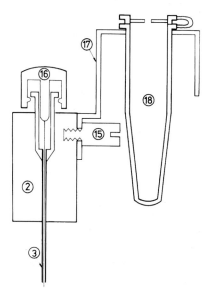

Figure 7 Sagittal cross-section of plexiglass headpiece (2) with fixed stand (17) for two polyethylene tubes (18).

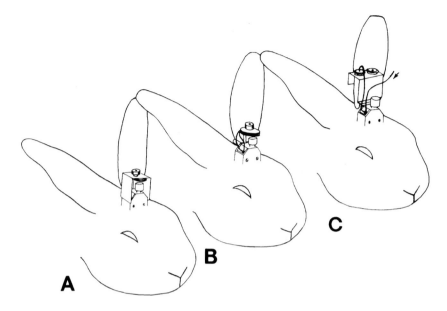

Figure 8 Rabbits with implanted headpiece (1) and microdialysis probe with cover (A), after removal of the cover (B) and during collection of the outflowing dialysate (C).

guide cannulae and the pistons fixed by screws. During insertion of the microdialysis probe it was continously perfused with 0.9% NaCl solution at 1.0 μl per min. During subsequent days after probe implantation microdialysis was performed at least twice weekly for several hours.

Samples of the outflowing fluid were collected in the polyethylene tube with 20 μl 1 N acetic acid and kept in the aluminium stand fixed to the rear side of the headpiece. The outflowing fluid was collected sequentially during 2- or 3-h periods. After measuring the volume of the collected samples they were lyophilized and maintained in this condition until radioimmunoassayed.

GnRH Radioimmunoassay

The concentration of gonadotropin-releasing hormone (GnRH) was measured in the hypothalamic microdialysate using antiserum No 6043-8 (donated by Prof. B. Kerdelhué, Unite de Neurobiologie de la Reproduction, Centre National de la Rechersche Scientifique, Jouy-en-Josas, France). The specificity of the antiserum was described in an earlier paper (Kerdelhué *et al.*, 1976) and used at a final dilution of 1:20,000. Luteinizing hormone releasing hormone (LHRH; Serva, Cat. No. 52345) was used as a reference standard and ligand for iodination. The intra- and inter-assay variations were 2.5% and 22.4% respectively. Tracer displacement (50%) was produced by about 26.6 pg of GnRH. The lyophilized hypothalamic microdialysates were redissolved in soluble distilled water to the initial volume and GnRH was determined by radioimmunoassay.

Three months after microdialysis probe insertion its position in the hypothalamus was marked by perfusion of 10 μl of a 10% Lissamine Green B (Sigma) solution through the probe followed by 30 μl of a 0.9% NaCl solution. Later the animals were sacrificed with a lethal dose of hexobarbital-natrium (Germed), the heads were perfused with isotonic saline followed by a 10% formalin solution, cut off and immersed in the 10% formalin solution for one week.

Histological Verification of the Position of Microdialysis Probe Tips

The block containing the hypothalamus and neigbouring structures was cut out from the rabbit's brain, washed in tap water for 24 h and then dehydrated in ethanol, cleared in methyl salicylate and embedded in paraffin wax. The time of processing was longer than in a routine histology. Serial coronal sections at 45 μm thick were cut from the paraffin block on a rotary microtome. The sections were flattened on hot water (45 °C) and picked up on gelatin-coated slides. After drying the paraffin was removed by immersing the sections four times in xylene; they were then mounted in DePeX, i.e. neutral solution of polystyrene and plasticizers in xylene (Serva, Cat. No. 18243) under a coverslip.

At the stage of methyl salicylate clearing, excellent visualisation of the position of the tip of microdialysis probe was noted. This was caused by the presence of Lissamine Green in the brain tissue. In some cases the polycarbonate membrane of the microdialysis probe tips remained in the brain tissue after the removal of the probe. In these cases the membrane stained by the Lissamine Green was very visible, which helped in the selection of the slices for evaluation of the brain area in which microdialysis took place.

Microdialysis Probe

The microdialysis probes used for hypothalamic, caudate nucleus and hippocampal implantation were manufactured by CMA/Microdialysis AB. Those used for caudate nucleus and hippocampus had a steel shaft with O.D. of 0.64 mm and a length of 20 mm (Cat. No. 8309504), membrane O.D. of 0.5 mm and length of 4 mm. Those implanted into the hypothalamus had a length of 30 mm (Cat. No. 8309540), or of 20 mm (Cat. No. 8309504). The polycarbonate membrane at the tip had O.D. of 0.5 mm and 3 or 4 mm length, with wall thickness of 60 μm and molecular cut off approximately 20,000 Daltons. The outlet tubing was cut shorter to a length of 100 mm, that is 1/3 of the inlet tubing length. Perfusion was performed with degassed 0.9% NaCl at 1 μl per min using 1 ml syringe and a CMA/Microdialysis AB 100 Microinjection Pump.

The microdialysis probes were fixed in the plexiglass pistons with O.D. of 4 mm and 5–8 mm height depending on how deeply the tips of the probe should be inserted into the subcortical structures.

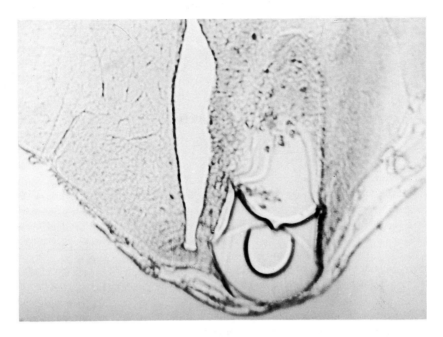

Figure 9 Coronal section of the hypothalamus of rabbit No. 1382 with the microdialysis probe tip. Magnification × 60.

RESULTS

From more than ten operated animals with implanted headpieces and microdialysis probes only two maintained the efficacy of the probes three months after their insertion into the hypothalamus. In these animals microdialysis was performed twice weekly with degassed 0.9% NaCl solution to prevent air bubbles or crystal formation inside the probes. The loss in efficacy occurred much earlier in the remaining animals with microdialysis performed twice weekly, not only with 0.9% NaCl solution, but also with artificial cerebrospinal fluid. The cause was probably a crystal formation inside the probe from the perfused artificial cerebrospinal fluid. Successful microdialysis was performed in two bilaterally ovariectomized rabbits on the 7th day after the insertion of the probes into the hypothalamus. The highest and the lowest values obtained on the day after probe insertion were 2.23 pg and 10.8 pg of GnRH per hour respectively.

In the rabbit (No. 1382) in which the position of microdialysis probe tip was visible on the coronal section of the hypothalamus (Figure 9), that is in the region of the infundibular group of perikarya of GnRH neurones (Barry, 1976), the microdialysis was performed 22 times. The first 6 h of microdialysis was performed on the 5th day and the last one on the 86th day after the probe implantation. The amount of GnRH

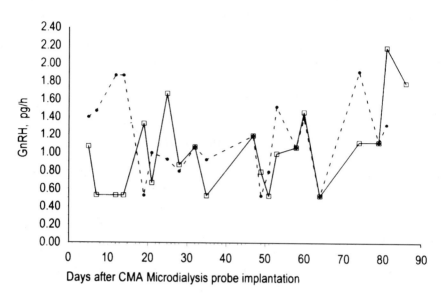

Figure 10 Release of GnRH from hypothalamus of rabbit No. 1382 into microdialysate fluid between the 5th and 86th day after microdialysis probe implantation. Open squares indicate the amount of GnRH/h in microdialysate in the first samples (1–180 min) and black circles in the second samples (181–360 min).

released into microdialysate varied according to the day, but a tendency towards an increase or decrease over the period of 81 days was not noted (Figure 10).

Looking at the amount of GnRH released into the first (1–180 min) and the second (181–360 min) samples, there was no specific correlation between them. The amount of GnRH in the first samples was equal to the amount of GnRH in the seconds, on 5 microperfusion days. The amount of GnRH was highest in the first samples when compared to the second samples. This was also observed on 7 microperfusion days. The results were opposite on 8 microperfusion days that is the amount of GnRH in the second sample was higher compared to the first sample.

CONCLUDING REMARKS

In one report (Pau and Spies, 1986), the mean level of GnRH in the push-pull perfusate obtained in ovariectomized rabbits one week after push-pull cannula implantation the amount of GnRH in the perfusion fluid was approximately 3.6 (2.5–5.0) pg/ml. In another report (Pau et al., 1986), individual GnRH episodes with pulse amplitude ranging between 1.5–12.5 pg/ml were observed in ovariectomized rabbits. Comparison of the values obtained by microdialysis and push-pull perfusion

(Pau *et al.*, 1986; Pau and Spies, 1986) indicates that the membrane is very permeable to GnRHs, and with low flow and filtration rate (short outflow tubing) the results are quite comparable.

A much lower difference can be expected in the amount of GnRH release in the successive samples on the same experimental day, or between successive microdialysis when compared with the data obtained with the push-pull technique. The present data do not support these assumptions. The differences in GnRH release between particular microdialysis days are remarkable. At present we do not have any explanation for these results, especially when taking the fact into the account that these experiments were performed on ovariectomized animals.

The present data on the GnRH release from the hypothalamus do not support another assumption that the amount of GnRH released from the hypothalamus would differ significantly according to the time which elapsed after the probe implantation. It would seem that the amount of GnRH released is independent of the reactive gliosis induced by the implantation of a microdialysis probe.

The loss of efficacy of the microdialysis probes some weeks after the implantation into subcortical structures may be associated with the construction of CMA/microdialysis AB probes. It is likely that the low internal diameter of the metal inlet tubing plays a role in probe blockage.

ACKNOWLEDGEMENTS

We thank Mrs. Krystyna Sadzinska for her help with the surgery, Mrs. Anna Kliszko for her excellent technical assistance and for drawing the figures. We are grateful to Mr. Aleksy Sobczyk for manufactering the rabbit's headpieces and rabbit's experimental boxes. We are indebted to Professor Bernard Kerdelhué (Unite de Neurobiologie de la Reproduction, Centre National de la Rechersche Scientifique, Jouy-en-Josas, France), for providing the highly specific antiserum to GnRH, Dr Andrzej Godlewski (Dept. Histology, Medical University of Lodz, Poland) for rabbit brain histology, and Dr Harold G. Spies (Oregon Regional Primate Research Center, USA) for reading the manuscript and for valuable remarks.

This study was supported by Medical University of Lodz research fund, and partially by the European Community (BIOMED-1, Associated Contract ERBBMHICT 921193).

REFERENCES

Barry, J. (1976) Characterization and topography of LH-RH neurons in the rabbit. *Neurosci. Lett.*, **2**, 201–205.

Kerdelhué, B., Catin, S., Kordon, C. and Jutisz, M. (1976) Delayed effects of *in vivo* immunoneutralization on gonadotropins and prolactin secretion in the female rat. *Endocrinology*, **98**, 1539–1549.

Kitamura, T., Tsuchihashi, Y. and Fujita, S. (1978) Initial response of silver-impregnated 'resting microglia' to stab wounding in rabbit hippocampus. *Acta Neuropathol. (Berl.)*, **44**, 31–39.

Morshead, C.M. and van der Kooy, D. (1990) Separate blood and brain origins of proliferating cells during gliosis in adult brains. *Brain Res.*, **535**, 237–244.

Pau, K.-Y.F. and Spies, H.G. (1986) Estrogen-dependent effects of norepinephrine on hypothalamic gonadotropin-releasing hormone release in the rabbit. *Brain Res.*, **399**, 15–23.

Pau, K.-Y.F., Orstead, K.M., Hess, D.L. and Spies, H.G. (1986) Feedback effects of ovarian steroids on the hypothalamic-hypophyseal axis in the rabbit. *Biol. Reprod.*, **35**, 1009–1023.

Sawyer, C.H., Everett, J.W. and Green, J.D. (1954) The rabbit diencephalon in stereotaxic coordinates. *J. Comp. Neurol.*, **101**, 801–824.

Traczyk, W.Z. (1966) The vasopressin content in perfusion fluid from the hypothalamus in conscious chronic dogs. *Bull. Acad. Pol. Sci. Ser. Sci. Biol.*, **14**, 727–729.

Traczyk, W.Z., Pau, K.-Y.F., Kaynard, A.H. and Spies, H.G. (1992) Effects of substance P or noradrenaline administration on hypothalamic GnRH and pituitary gonadotropin secretion in conscious rabbits. First International Congress of the Polish Neuroscience Society. Abstracts. Warsaw, 21–23 September 1992. *Acta Neurobiol. Exp.*, **52**, 194.

Numbers in Figures

1 Plexiglass headpiece implanted into rabbit's skull bone for caudate nucleus and hippocampal microdialysis.

2 Plexiglass headpiece implanted into rabbit's skull bone for hypothalamus microdialysis.

3 3rd cerebral ventricle guide cannula.

4 Stilette in the piston introduced into the guide cannula.

5 Screw fixing the piston with the stilette or with microdialysis probe.

6 Microdialysis probe in the piston for caudate nuclei microdialysis.

7 Microdialysis probe in the piston for hippocampus microdialysis.

8 Microdialysis probe with the tip membrane in caudate nuclei.

9 Microdialysis probe with the tip membrane in hippocampi.

10 Microdialysis probe with the tip membrane in hypothalamus.

11 Nut fixing the aluminium cover.

12 Aluminium cover of headpiece.

13 Coil for inlet and outlet tubings.

14 Stand for coil.

15 Screw fixing the stand for coil.

16 Closure cap with stilette fixed in the 3rd cerebral ventricle cannula.

17 Aluminium stand for two polyethylene tubes for outflow fluid collection, fixed to the headpiece.

18 Polyethylene tubes for outflow fluid collection.

PART C
IN VIVO: VOLTAMMETRY AND MICROCAPSULES

I. INTRODUCTION

E.C.M. de LANGE AND A.G. de BOER

Division of Pharmacology, Leiden/Amsterdam Center for Drug Research (LACDR), University of Leiden, P.O. Box 9503, 2300 RA Leiden, The Netherlands

In vivo voltammetry can be used to measure signals arising from electrochemically active species present near the electrode. In relation to other *in vivo* techniques an important aspect of *in vivo* voltammetry is its speed. Thus, especially highly dynamical processes can be investigated as a function of time (msec-scale) in a very specific area in the brain. For this reason it may be a useful complement to other *in vivo* techniques.

In the complex medium of the brain only the easily oxidizable compounds present in the extracellular fluid will contribute to the signal (current). However, because no separation step takes place before oxidation of compounds occurs, selectivity is a major problem in the use of this technique. Tissue reactions due to the implantation of the electrode will be less than after implantation of a microdialysis probe, because of the significantly smaller size of the electrodes.

A number of voltammetric techniques are applied to increase selectivity. Using the differential pulse mode (DVP), compounds that oxidize at different potentials will show up as separate peaks if their oxidation potentials are sufficiently different. With cyclic voltammetry (CV) diagnostic information about electrochemical reaction can be provided. Also the use of electrodes with specific sensitivities may provide more information about the reacting compound. Nevertheless care should be taken in the interpretation of the relation between current and the concentration of the compound of interest.

In the third part of this book a study on the use of liposomes as a drug carrier to transport drugs to the CNS is presented.

II. URIC ACID PASSAGE ACROSS BLOOD-BRAIN BARRIER TO THE CEREBRAL CORTEX AND CORPUS STRIATUM OF THE RAT AS MONITORED WITH DIFFERENTIAL PULSE VOLTAMMETRY (DPV) AND FAST SCAN CYCLIC VOLTAMMETRY (FCV)

J. PAVLÁSEK, M. HABURČÁK AND C. MAŠÁNOVÁ

Department of Neurophysiology, Institute of Normal and Pathological Physiology,
Slovak Academy of Sciences, Sienkiewiczova 1, 81371 Bratislava, Slovak Republic

Differential pulse voltammetry (DPV) and fast scan cyclic voltammetry (FCV) with a carbon fibre microelectrode was used in pentobarbital anaesthetized rats for monitoring the passage of the uric acid (UA) across blood-brain barrier (BBB) into the cerebral cortex (CC) and corpus striatum (CS). Elevation of the UA plasma concentration (approximately up to 0.5 mM) by UA injection (during the period of 2–3.5 min, 1 ml, about 12 mM) into the internal carotid artery resulted in increased UA level in the brain. Voltammetric signal (DPV) representing UA redox current reached its maximum within five min after the UA injection: 4.2 ± 1.5 min in the CC and 4.3 ± 1.2 min in the CS ($n = 5$, means $\pm$ SD). The onset of the increase in UA level in the brain (CS) was detected (FCV) 100 s after UA injection had started. According to the calibration curve obtained *in vitro* (FCV) the maximal value of the UA redox current observed in the CS (FCV) corresponded approximately to 0.2 mM UA. The lowering of the UA signal was not as equally prompt as its raising phase; at the end of the experiments, about 59 min after UA injection, the mean of the UA redox current (DPV) was 48% (CS) and 28% (CC) of its maximal value. The UA passage across BBB is one of the mechanisms helping the brain tissue to cope with a radical and oxidant load and to maintain a cerebral homeostasis.

INTRODUCTION

Uric acid (UA) is the final product of the enzymatic catabolism of adenine- and guanine-based purine compounds in man and higher primates; other mammals excrete allantoin and urea as the major nitrogen-containing degradation of purines.

An important attribute of the UA is that it is contained in all extracellular fluid compartments — among others also in the cerebrospinal fluid (Becker, 1993). Production of the UA in the brain tissue or its transport across blood-brain barrier (BBB) to the extracellular space of the brain is not yet satisfactorily understood. It is supposed that the extracellular brain UA may simply correlate with the energy use (amount of metabolized energy-related purines like ATP, GTP) or provide a marker for the adenosine release in the process of the 'purinergic' synaptic transmission (O'Neill, 1986; Mueller and Kunko, 1990). As for UA passage across the layer of endothelial cells in the brain capillaries some results have indicated that UA cell membrane transport could be relatively slow (Lassen, 1961); this finding supports the existence of the UA concentration gradient between plasma and cerebrospinal fluid (approximate range in man 140–440 μmol·l^{-1} versus 20–80 μmol·l^{-1}).

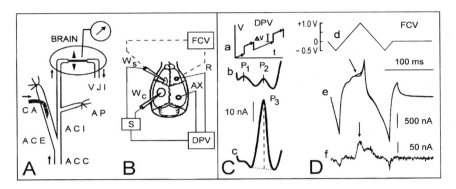

Figure 1 (A) Schematic diagram of cannulation of the external carotid artery (ACE) enabling drug administration into the rat brain via the internal carotid artery (ACI) without interruption of the blood flow in the common carotid artery (ACC) and ACI. (AP – pterygopalatine artery, VJI – internal vena jugularis, CA – cannula). Scheme of measuring device for voltammetric recordings. (B) Experimental setup. Placement of electrodes (R – reference, AX – auxiliary, W_s/W_c – working electrode positioned in corpus striatum/cortex, S – switch between W_s and W_c electrodes) and devices used for differential pulse voltammetry (DPV) and fast scan cyclic voltammetry (FCV). (C) Differential pulse voltammetry (DPV) and DPV voltammograms. (a). Waveform of the potential (V) applied to the W electrode at time t. Two measurements (dots) are made for each pulse (ΔV) and the differential current is plotted against the applied potential to obtain the voltammogram; (b). Control DPV voltammogram recorded *in vivo* from the cortex before uric acid (UA) administration. Maximum amplitudes of the peak P_1 and P_2 were reached at 50 mV and 440 mV respectively; (c). Formation of the peak P_3 at 430 mV 3.3 min after injection of the UA into the ACI. The height of the peaks was measured as indicated by the dashed line. (D) Fast scan cyclic voltammetry (FCV) and FCV voltammograms; (d). Voltage waveform used for FCV applied to the W electrode; (e). Current waveforms observed in the corpus striatum before and after UA administration. Two waveforms are superimposed. The resultant current increment in UA is shown by arrow. The height of the peak is used to quantify the amount of UA present in the vicinity of the W electrode; (f). Current-time fast cyclic voltammogram. Amplified signal produced by subtraction of the two waveforms shown in (e) is denoted by an arrow.

The increase in the interest for the origin and function of the extracellular brain UA is motivated by cumulation of evidence of a beneficial physiological action of the UA in connection with its antioxidant capacity and radical scavenging properties (Becker, 1993). These attributes of the UA can play a very important role in the protection of the nervous tissue against a damage caused by an oxygen stress and free radicals formation (Rose and Bode, 1993) by xanthine oxidase (Tan *et al.*, 1993) under the conditions of ischemia-reperfusion episodes (Kvaltínová *et al.*, 1993) occurring in the course of the brain stroke.

The voltammetric technique described below is employed for the investigation of UA transport across BBB occurring in parallel with the elevation of the UA plasma levels.

MATERIALS AND METHODS

The experiments were carried out on male Wistar rats with an average body weight (BW) of 380 g and kept on Larsen diet. Animals were anaesthetized with pentobarbital (Spofa, Prague), 5% solution in physiological saline, 0.1 ml/100 g BW, i.p.; about one third of this dose was added after approximately 40 min (duration of the experiment usually did not exceed 120 min).

A cannula was placed into the external carotid artery (Figure 1A); this technique (Ježová et al., 1989) permits the application of solution into the internal carotid artery without obstructing its free blood flow.

Animals were fixed in a stereotaxic apparatus and four small openings were drilled in the skull (Figure 1B) for voltammetric electrodes:

(a) Working electrode (glass carbon fibre electrode) — (Pavlásek et al., 1993), treated electrochemically as described in Mermet and Gonon (1988) in the left parietal cortex (W_c) with stereotaxic coordinates (Fifková and Maršala, 1960): AP +3.0, L +2.5, V +1.5;
(b) Working electrode in the left corpus striatum (W_s) with coordinates: AP −1.0, L +2.0, V +3.5 to +4.5;
(c) Auxiliary (AX) electrode (a stainless steel watchmaker's screw) positioned in the parietal region of the right hemisphere.
(d) Reference (R) electrode (Ag/AgCl wire) in the frontal region of the right hemisphere. The incisions in the dura mater were made for the W_c and W_s electrodes; AX and R electrodes were placed epidurally.

Differential pulse voltammetry — DPV (Justice, 1987) or fast-scan cyclic voltammetry — FCV (Millar and Barnett, 1988) were used for recording of electrochemical signals.

Polarographic analyzer for DPV (PA 4, Laboratory Equipment, Prague) worked with a three-electrode system (Figure 1B) with the following parameters: speed of the linear potential sweep 100 mV·s^{-1}, potential limits from −100 mV to +800 mV, pulse amplitude 50 mV, pulse duration 60 ms (current sampling 20 ms before the pulse and again 20 ms before the end of the pulse), pulse period 0.2 s (Figure 1C, a). The voltammetric signal was drawn with an x-y plotter (XY 4106, Laboratory Equipment, Prague). Switch (S) enabled the selection between W_c and W_s (Figure 1B) and performance of parallel monitoring of the chemical changes in the micro-environment of the cortex and striatum. The interval between consecutive voltammetric recordings with W_c – W_s electrode was 1 min.

FCV measured using an IBM PC (software of the Institute of Normal and Pathological Physiology, Bratislava) employed the following potential waveform applied to the W_s electrode: 1.5 cycles of 10 Hz triangular ramp, scanning between −0.5 and +1.0 V relative to the R electrode at 30 V/s, with zero potential between scans (Figure 1D, d). Interscan intervals were 20 s. Signal was amplified by an I/V converter (Institute of Normal and Pathological Physiology, Bratislava) and interfaced to the computer with a digital input/output board (PC-LabCard PCL-812, Taiwan) for waveform capture, storage and analysis.

Uric acid — UA (Loba Feinchemie) was dissolved in glycerol (Lachema, Brno) and to one part of this mixture ten equal parts of saline were added; the UA solution was always freshly prepared and filtered prior to its administration. Concentration of the UA solution varied (about 12 mM) because in every experiment, in spite of the differences in animals' BW (w, in grams), the same UA concentration in the blood was to be attained (0.5 mM) while keeping the injected volume constant (1 ml). The formula used for the calculation of the blood volume (V) was as follows (Lee and Blaufox, 1985):

$$V(\text{ml}) = (w * 0.06) + 0.77$$

UA injection (during the period of 2 to 3.5 min) was followed by instillation of 0.2 ml saline with 100 units of heparin.

When voltammetric signals became stabilized (10–15 recordings for DPV or FCV), UA injection started and multiple-time point voltammetric measurements were made.

The quantification of the electrochemical signals recorded with FCV (Millar *et al.*, 1985, Figure 1D, f) or DPV (Figure 1C, b and c) was carried out by measuring the amplitude of the peaks representing redox current.

The Student's *t*-test was used to evaluate the results (mean values and standard deviations are shown).

RESULTS

On the control voltammetric (DPV) recordings (Figure 1C, a) — prior to injection of the UA — two clearly identifiable peaks were present (Figure 1C, b). The first one (P_1) formed at the lower voltage corresponds to ascorbic acid (Lane *et al.*, 1976; Pavlásek *et al.*, 1994), while the other (P_2) occurring at the voltage in the range from 360 to 500 mV represents redox current from catecholamines and their metabolites (CA.OC) as previously reported by Lane *et al.* (1976) and Pavlásek *et al.* (1992).

UA administration caused an increase in the peak P_2 which occurred in the two studied structures of the brain. As determined by *in vitro* calibrations, the peak of the UA redox current ranged from 310 to 390 mV (4 measurements). Therefore it is reasonable to suppose that with the type of the W electrode and technique of UA injection used, this augmentation was the result of the overlapping of the electrochemical signals (Guadalupe *et al.*, 1992) representing CA.OC and UA redox current; this complex peak was denoted as P_3 (Figure 1C, c).

The time-course of the rise of the peak P_3 gave information about the dynamics of the UA transport across BBB in the striatum and in the cortex. In each experiment the mean of the five consecutive measurements directly preceding UA injection served as the control (100%) as shown in Figure 2.

In the cortex peak P_3 attained its maximal value within five min after the UA injection (4.2 ± 1.5 min; the number of experiment $n = 5$). This maximal value of the peak P_3 expressed in per cent of the CA.OC control ($P_2 = 100\%$) was $797 \pm 285\%$

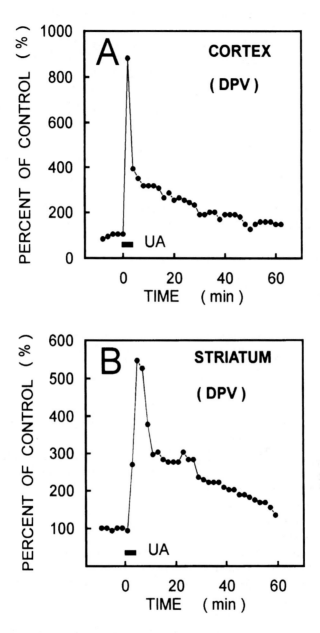

Figure 2 Time course of the changes of uric acid (UA) redox current (peak P_3) after UA injection (horizontal bar denotes administration period) in cortex (A) and corpus striatum (B) of the rat measured by DPV. Results are expressed as the percentage of the mean of the control CA.OC values. The amplitude of mean control CA.OC (peak P_2 before UA administration) was set at 100%. Results from one experiment.

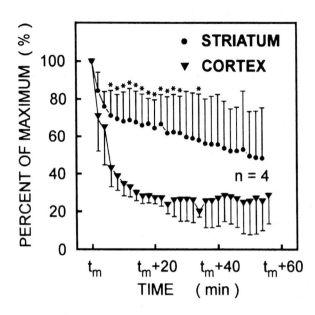

Figure 3 Time course of the changes of the uric acid (UA) redox current (peak P_3) after UA maximum current was attained in the rat corpus striatum (dots) and in the cortex (triangles) measured by DPV. Ordinate: percentage of the UA maximum current (100%) attained at time t_m. Abscissa: time after UA maximum redox current was attained. n = number of experiments; asterisks indicate significant difference ($P < 0.05$) between values in corpus striatum and cortex.

($n = 5$, $P < 0.001$) as shown in Figure 2A. After the maximal size (100%) of the peak P_3 had been attained (at time t_m), its amplitude diminished (Figure 3). At the fourth minute after the peak P_3 reached its maximum ($t_m + 4$ min) its values were significantly lower than in the time t_m. Six minutes after its maximum ($t_m + 6$ min) peak P_3 lowered to $43 \pm 10\%$ ($n = 4$).

In the striatum peak P_3 also reached its maximal value within five min after the UA injection (4.3 ± 1.2 min, $n = 5$). This maximal value of the peak P_3 expressed in per cent of the CA.OC control ($P_2 = 100\%$) was $436 \pm 251\%$ ($n = 5$, $P < 0.02$) as shown in Figure 2B. After the maximal amplitude (100%) of the peak P_3 had been attained (at time t_m), its size diminished (Figure 3). At the second minute after the peak P_3 reached its maximum ($t_m + 2$ min) its values were significantly lower than at time t_m. Six min after its maximal level ($t_m + 6$ min) peak P_3 dropped to $71 \pm 13\%$ ($n = 4$).

There was no significant difference ($P > 0.05$) between the time periods in which peak P_3 attained its maximum after the beginning of the UA injection in the striatum and in the cortex. However, following the maximum of the peak P_3, significantly ($P < 0.05$) lower values of the peak P_3 in the cortex were observed during the time period from $t_m + 6$ min to $t_m + 28$ min i.e. in the diminishing phase (Figure 3). After $t_m + 36$ min the differences in the values of the peak P_3 in the cortex and in the striatum were insignificant.

Fast cyclic voltammetry (FCV, Figure 1D, d) enabled us to determine more exactly the dynamics of the initial phase of the UA passage across BBB. The sensitivity of the FCV method does not allow measurement of basal (background) concentration; it is best equipped to record changes in levels (Stamford, 1990) of the agents which can readily cross the BBB (Figure 1D, e). After subtraction of the background signal (charging current, lower recording) from the signal recorded after UA administration (Faraday current, denoted by an arrow), a difference in the FCV signal (Figure 1D, f) represented the concentration change which resulted from UA injection. Oxidation peak of the UA (Figure 1D, f, denoted by arrow) occurred at polarization voltage of 780 mV.

Dynamics of the UA transport into the brain monitored by FCV is shown in Figure 4A. Approximately 100 sec after the UA application started (about 80% of the whole UA dosis were injected) the signal representing UA redox current occurred as shown by the differences in FCV recordings. In the course of the following 100 sec there was an almost linear increase in this current up to the value of 87 nA (Figure 4A) which was close to the maximal UA redox current of 91 nA attained in this experiment 280 sec after the UA application had started. This result confirmed observations performed using DPV — in the striatum the maximum of the UA redox current occurred within five min after UA administration. After reaching the maximum level the UA redox current diminished: 6 min later it lowered to 49% and after the next 10 min to 30% of its highest value.

In order to assess the UA concentration in the striatum an *in vitro* calibration with FCV technique was made (Figure 4B) using the same W electrode as in the experiment illustrated in Figure 4A. According to the calibration curve, the maximal value of the current (91 nA) observed in the experiment shown in Figure 4A corresponded approximately to 0.2 mM UA.

DISCUSSION

In the control group of untreated rats ($n = 7$) UA plasma levels were measured with an enzymatic method (Dept. of Clinical Biochemistry, Academician Derer's State Hospital, Bratislava); values ranged from 15 μM to 121 μM (55 $\pm$ 33 μM). As determined in four *in vitro* measurements (UA solute added to a known volume of the blood) the sensitivity of this enzymatic method was 17 $\pm$ 11%.

In five experiments, small samples of the blood were collected (via carotid artery) 4, 5, 8, 12 and 13 min after UA injection (1 ml, about 12 mM, equivalent to UA plasma concentration of approximately 0.5 mM — calculated in relation to the estimated blood volume) and UA plasma concentration was enzymatically measured in these samples. There were large differences in UA concentration between specimens taken at shorter intervals after UA injection (4, 5, 8 min) and samples collected at longer

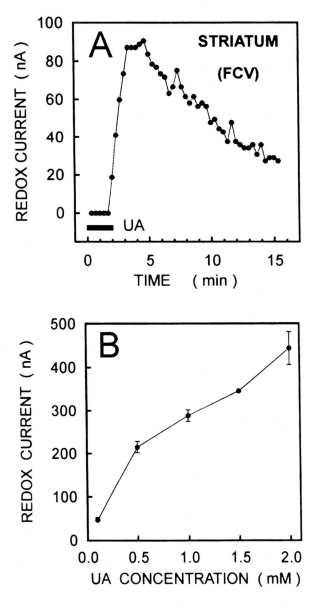

Figure 4 Uric acid (UA) redox current measured by fast scan cyclic voltammetry (FCV). (A) Time course of the changes of UA redox current after UA injection (horizontal bar denotes administration period) in the rat striatum. Results from one experiment. (B) *In vitro* calibration curve relating the UA redox current to the concentration of UA measured with W electrode immersed in solution with different UA concentration in saline. The calibration was made with the same electrode used in the *in vivo* experiment illustrated in Figure 4A. Results are means ± S.D. from 2–10 scans for each UA concentration.

intervals (12 and 13 min). In the first group values were 291 μM, 538 μM and 513 μM respectively, while in the second group (with longer intervals) they were much lower — 47 μM and 37 μM respectively. This rapid decrease in UA plasma concentration to the level of the control values indicated an access of UA to the extracellular and intracellular compartments in the other tissues (Becker, 1993) as well as the effectiveness of the processes resulting in UA metabolization by enzymatic and by non-enzymatic means (Kaur and Halliwell, 1990; Sorensen, 1960).

In one of the observed brain structures (striatum) changes in the voltammetric signal (FCV), reflecting UA redox current, were detected about one and a half minute after UA administration. The value of *in vivo* voltammetry lies in its unique ability to provide 'on line' measurement i.e., nearly continuous measurement of the concentration changes of the electroactive compounds in the extracellular space of the brain with discrete measurement probes enabling high spatial resolution and causing only little perturbation of the system during the measurement. The electro-chemistry of UA at the pyrolytic graphite electrode has been studied extensively by Dryhurst (1972). *In vitro* testing of the W electrodes used in our laboratory confirmed their sensitivity for UA in the DPV mode in the range of tens of μM.

In the time period during which high UA plasma levels were measured (4–8 min after the UA injection) the amplitudes of the peak P_3 ranged from 105% to 796% (in the striatum), and from 275% to 1166% (in the cortex) of the control (peak P_2). Maximal amplitudes of the peak P_3 were observed 3 to 5 min after UA injection had started in the striatum and 2 to 6 min in the cortex (DPV; please notice $W_s - W_c$ switching in one min interval). As shown by *in vitro* calibration of the UA signal measured *in vivo* (FCV), maximal UA redox current observed in the striatum corresponded approximately to 200 μM UA.

Rapid decrease in the UA plasma levels to the control values (12–13 min after UA administration) was not accompanied by an equally prompt lowering of the peak P_3; at the end of the experiments, about 59 min after UA injection, the mean of the peak P_3 was 48% (striatum) and 28% (cortex) of its maximum (approximately 210% and 220% of the control represented by peak P_2).

There was only little regional heterogeneity in the dynamics of the UA signal changes between cortex and striatum. UA passage into both structures was equally rapid; the significant difference was observed only in decaying phase of the UA signal, following the attainment of its maximum level. This phenomenon may be explained by differences in the density of the capillaries and local blood flow as well as by all anatomical, physicochemical and biochemical mechanisms regulating metabolism and transport processes across cell membranes.

It has been reported that UA, and its monovalent anion urate, play a role of an effective non-enzymatic antioxidants: their scavenging activity was estimated to be 30–65% (for peroxyl radicals) and 10–15% (for hydroxyl radicals) of the whole scavenging capacity of all other plasma constituents (Becker, 1993). It is therefore biologically relevant to consider UA passage across BBB as one of the mechanisms helping the brain tissue to cope with a radical and oxidant load and to maintain a cerebral homeostasis.

CONCLUDING REMARKS

The passage of uric acid (UA) across the blood-brain barrier (BBB) is one of the mechanisms that is essential for the brain tissue to cope with a radical and oxidant load, and to maintain a cerebral homeostasis. Two voltammetric techniques, namely Differential Pulse Voltammetry (DPV) and Fast-scan Cyclic Voltammetry (FCV) have been successfully used to monitor UA passage across the BBB.

ACKNOWLEDGEMENTS

This work was supported, in part, by Grant agency for science (Grant 2/999323). The financial support of the European Community (BMH1-CT92-1193, PECO Associated Contract #ERBBMHICT 921193) is gratefully acknowledged.

REFERENCES

Becker, B.F. (1993) Towards the physiological function of uric acid. *Free Radic. Biol. Med.*, **14**, 615–631.

Dryhurst, G. (1972) Electrochemical oxidation of uric acid and xanthine at the pyrolytic graphite electrode. *J. Electrochem. Soc.*, **119**, 1659–1664.

Fifková, E. and Maršala, J. (1960) Stereotaxic atlas for the cat, rabbit and rat brains. *Avicenum*, pp. 41–58. Praha.

Guadalupe, T., Gonzales-Mora, J.L., Fumero, B. and Mas, M. (1992) Voltammetric monitoring of brain extracellular levels of serotonin, 5-hydroxyindoleacetic acid and uric acid as assessed by simultaneous microdialysis. *J. Neurosci. Methods*, **45**, 159–164.

Ježová, D., Johansson, B.B., Opršalová, Z. and Vigaš, M. (1989) Changes in blood-brain barrier function modify the neuroendocrine response to circulating substances. *Neuroendocrinology*, **49**, 428–433.

Justice, J.B. Jr. (1987) Introduction to *in vivo* voltammetry. In: Justice, J.B. Jr. (Ed.), *Voltammetry in the Neurosciences*, pp. 3–101. Clifton, New Jersey: The Humana Press.

Kaur, H. and Halliwell, B. (1990) Action of biologically-relevant oxidizing species upon uric acid. Identification of uric acid oxidation products. *Chem. Biol. Interact.*, **73**, 235–247.

Kvaltínová, Z., Lukovič, L. and Štolc, S. (1993) Effect of incomplete ischemia and reperfusion of the rat brain on the density and affinity of α-adrenergic binding sites in the cerebral cortex. Prevention of changes by stobadine and vitamin E. *Neuropharmacology*, **32**(8), 785–791.

Lane, R.F., Hubbard, A.T., Fukunaga, K. and Blauchard, R.J. (1976) Brain catecholamines: Detection *in vivo* by means of differential pulse voltammetry at surface-modified platinum electrodes. *Brain Res.*, **114**, 346–352.

Lassen, U.V. (1961) Kinetics of uric acid transport in human erythrocytes. *Biochim. Biophys. Acta*, **53**, 557–569.

Lee, H.B. and Blaufox, M.D. (1985) Blood volume in the rat. *J. Nucl. Med.*, **25**, 72–76.

Mermet, C.C. and Gonon, F.G. (1988) Ether stress stimulates noradrenaline release in the hypothalamic paraventricular nucleus. *Neuroendocrinology*, **47**, 75–82.

Millar, J., Stamford, J.A., Kruk, Z.L. and Wightman, R.M. (1985) Electrochemical, pharmacological and electrophysiological evidence of rapid dopamine release and removal in the rat caudate nucleus following electrical stimulation of the median forebrain bundle. *Eur. J. Pharmacol.*, **109**, 341–348.

Millar, J. and Barnett, T.G. (1988) Basic instrumentattion for fast cyclic voltammetry. *J. Neurosci. Methods*, **25**, 91–95.

Müller, K. and Kunko, P.M. (1990) The effect of amphetamine and pilocarpine on the release of ascorbic and uric acid in several rat brain areas. *Pharmacol. Biochem. Behavior*, **35**, 871–876.

O'Neill, R.D. (1986) Adenosine modulation of striatal neurotransmitter release monitored *in vivo* using voltammetry. *Neurosci. Lett.*, **63**, 11–16.

Pavlásek, J., Haburčák, M., Mašánová, C. and Orlický, J. (1993) Increase in catecholamine content in the extracellular space of the rat's brain cortex during spreading depression wave as determined by voltammetry. *Brain Res.*, **628**, 145–148.

Pavlásek, J., Mašánová, C., Bielik, P. and Murgaš, K. (1992) Voltammetrically determined differences in changes evoked by KCl microinjections on catecholamine levels in the reticular formation and corpus striatum of the rat. *Physiol. Res.*, **41**, 191–200.

Pavlásek, J., Mašánová, C., Haburčák, M. and Murgaš, K. (1994) Catecholamine overflow within rat striatum: The influence of microstimulation and elecroconvulsive stimulation as observed with voltammetry. *Experimental Physiology*, **79**, 327–335.

Rose, R.C. and Bode, A.M. (1993) Biology of free radical scavengers: An evaluation of ascorbate. *FASEB J.*, **7**, 1135–1142.

Sorensen, L.B. (1960) The elimination of uric acid in man by means of ^{14}C-labelled uric acid. *Scand. J. Clin. Lab. Invest.*, **12** (suppl. 54), 12–14.

Stamford, J.A. (1990) Fast cyclic voltammetry: Measuring transmitter release in 'real time'. *J. Neurosci. Methods*, **34**, 67–72.

Tan, S., Radi, R., Gaudier, F., Evans, R.A., Rivera, A., Kirk, K.A. and Parks, D.A. (1993) Physiological levels of uric acid inhibit xanthine oxidase in human plasma. *Pediatr. Res.*, **34**, 303–307.

III. LIPOSOME MICROCAPSULES: AN EXPERIMENTAL MODEL FOR DRUG TRANSPORT ACROSS THE BLOOD-BRAIN BARRIER (BBB)

F. MIXICH AND S. MIHAILESCU

Faculty of Medicine Craiova, Dept. of Cell Biology and Physiology,
P.Rareş str. 4, 1100 Craiova, Romania

Experimental studies showed that liposomes (artificial lipid vesicles) can cross the wall of continuous capillaries in the myocardium and in the central nervous system (CNS). Liposomes can therefore be used as drug carriers in these tissues. Data on interactions between liposomes and endothelial cells, which could explain the passage of liposomes through the wall of continuous capillaries, have previously been reported. In our study in the rat, liposomes were injected intravascularly in both *in situ* and isolated heart experiments and their interactions with endothelial and parenchymal cells was studied with the use of electron microscopy. Liposomes were prepared by a reverse evaporation technique from phosphatidylcholine, cholesterol and diacetylphosphate. In isolated heart experiments liposomes were marked by encapsulating peroxidase inside them, in order to make a distinction between liposomes and naturally occurring membranes.

Both *in situ* and isolated heart experiments were performed in adult white rats. The results from *in situ* experiments show that endothelial cells from the aorta interact with liposomes by a process of phagocytosis. In isolated heart experiments liposomes were found, after 3 min of contact with the capillary endothelium, both in the interstitial space and inside myocardial cells.

The experimental procedures described in this chapter demonstrates the passage of liposomes through the wall of continuous capillaries by a process involving as first step phagocytosis. Liposomes may therefore be used as drug-carriers in tissues and organs with low capillary permeabilities, such as the cerebral circulation.

INTRODUCTION

The efficacy of free therapeutic agents is hindered by factors such as inability to direct them toward a specific tissue, potential antigenicity, and inactivation in the circulation. To overcome these difficulties, inherent in case of direct administration, several efforts have been made to find suitable carriers. At present, the most promising microcapsules that can be used in drug replacement therapy are liposomes, i.e., artificial lipid vesicles, which can entrap drugs within their aqueous compartment and/or lipid bilayer membranes (Bangham *et al.*, 1974; Ryman and Tyrell, 1979; Toffano and Bruni, 1980; Bankert *et al.*, 1985; Storm *et al.*, 1980).

Liposomes can be prepared by a variety of techniques that produce several different physical structures, ranging from the smallest unilamellar vesicles (SUV) of 20 to 50 nm in diameter, to much larger uni-, oligo-, or multi-lamellar vesicles of up to tens of microns in diameter (Bangham *et al.*, 1974; Szoka and Papahadjopoulos, 1978; Gregoriadis, 1984; Riaz *et al.*, 1988; Allen, 1994). The exact location of a drug in the liposome will depend on the drug's physicochemical characteristics and on the composition of the lipids (Fendler, 1980; Riaz *et al.*, 1988).

Experiments in several laboratories showed that mammalian cells *in vitro* and *in vivo* will incorporate large numbers of lipid vesicles without cytotoxic effects (Poste *et al.*, 1976; Pagano and Weinstein, 1978; Gregoriadis *et al.*, 1983; Leserman *et al.*, 1984). Despite extensive analysis of liposome-cell interactions, the mechanisms for cellular incorporation of liposomes and their contents are still uncertain, especially in cases of *in vivo* administration (Pagano and Weinstein, 1978; Weissman and Finkelstein, 1980; Poste, 1983). At ultrastructural level, it has been difficult to trace the path of liposome internalization due to the lack of a suitable marker (Poste, 1983; Waser *et al.*, 1987; Foldvari *et al.*, 1988; Nicolau and Cudd, 1989).

A considerable effort has been made to elucidate the fate of liposomes *in vivo* (Gregoriadis *et al.*, 1983; Poste, 1983). A large body of literature has accumulated over the last decade describing the tissue distribution of liposomes by various routes of administration (Gregoriadis *et al.*, 1983; Poste, 1983). The question of whether liposomes injected intravenously can cross blood vessels to gain access to extravascular tissues is of special importance. The answer to this question is fundamental in determining whether it may be feasible to target liposomes to a wide variety of cell types in the body. Surprisingly, very little research has been devoted to the study of this important topic. This, no doubt, reflects the substantial technical difficulties involved in studying transvascular transport processes *in situ*.

The anatomy of the microcirculation in different organs and tissues can be expected to be of crucial importance in determining whether liposomes can escape into the surrounding extravascular tissue. Viewed from a mechanical standpoint, continuous capillaries represent a major barrier for liposome extravasation. Many authors consider that liposomes are unable to leave the microcirculation in organs with continuous capillaries (skeletal, smooth and cardiac muscles, CNS, lung) (Poste, 1983). However, a large number of experiments demonstrate that liposomes with various encapsulated markers can leak their content into parenchymal cells after i.v. administration. There is, however, direct evidence for the mechanisms of interaction between liposomes and endothelial cells, which allows the delivery of substances through a capillary barrier (Zipper *et al.*, 1988).

Cerebral endothelium is of particular interest by virtue of its structural and metabolic regulation of blood-brain transport and alterations therein in a variety of disease states. The problem of transport of substances through the blood-brain barrier (BBB) by liposomes was extensively studied by Yagi and co-workers (Naoi and Yagi, 1980; Yagi *et al.*, 1982). The important finding arising from their study is that the addition of sulphatide (a glycolipid) to liposomes composition increases their ability to cross the BBB. Other authors (Nagai, 1982; Osanai and Nagai, 1984) confirmed these results by showing that the experimentally-induced allergic encephalomyelitis can be suppressed by a protease inhibitor, leupeptin, encapsulated in liposomes containing sulphatide.

The combination of sulphatide liposomes with monoclonal antibodies against glioma-associated antigens were predicted by Kito *et al.* (1987, 1989) as a liposomal drug delivery system for active targeting chemotherapy of glioma (carriers of third generation).

Other authors have shown that the uptake of liposome-entrapped drugs by the brain is either less effective than by other tissues (Hayakawa *et al.*, 1974a, 1974b; Takada *et al.*, 1981), or completely negative (Kito *et al.*, 1987). However, modifications of the liposome composition to achieve an increased transfer through the BBB are essential.

This chapter describes the experimental procedures for the use of liposomes as transport agents to cross the barrier represented by a continuous endothelium. The modalities of interaction of liposomes with endothelial and parenchymal cells are also described. Empty liposomes and liposomes marked with peroxidase were injected intravascularly in two experimental models: whole animals and isolated hearts. Liposomes interact with the endothelium and cross the wall of continuous capillaries, which justify their use as drug carriers across the BBB.

MATERIALS AND METHODS

Preparation of Liposomes

Preparation of 'empty' liposomes

Liposomes with no encapsulated marker ('empty' liposomes) were prepared by a reverse-phase evaporation procedure (REV) (Szoka and Papahadjopoulos, 1978). The schematic diagram of the system is shown in Figure 1. It consists essentially of components: a ultrasonicator (Ultrasonics Ltd. Rapidic, England), and a vacuum rotary evaporator.

REV liposomes were prepared from 26.4 μmole egg yolk phosphatidylcholine (Sigma, USA), 12.8 μmole cholesterol (Sigma, USA), and 4 μmole diacetylphosphate (Fluka, Switzerland). These lipids were dissolved in 5 ml diethylether, and 1.5 ml of phosphate-buffered saline (PBS) were added. The biphasic mixture composed of lipid-diethylether solution and aqueous buffer was sonicated for 60 sec at room temperature until a homogenous suspension of inverted micelles was obtained. The organic phase was then removed at 20 °C in a rotary evaporator, under vacuum. PBS (1.5 ml) was added to the resulting gel and this combination was stirred with a vortex mixer, for 10 minutes, at room temperature. The aqueous suspension of liposomes thus obtained was extruded through a Millipore filter with a pore size of 0.2 μm.

Preparation of 'marked' liposomes

REV liposomes with encapsulated horse-radish peroxidase as a marker, were prepared as described above, but the enzyme was added after sonication in order to avoid its inactivation. In brief, lipids, diethylether and PBS were sonicated for 60 sec, and peroxidase (Sigma) was added to the obtained emulsion resulting in a final concentration of enzyme of 1.5 mg/ml of final suspension of liposomes. After an

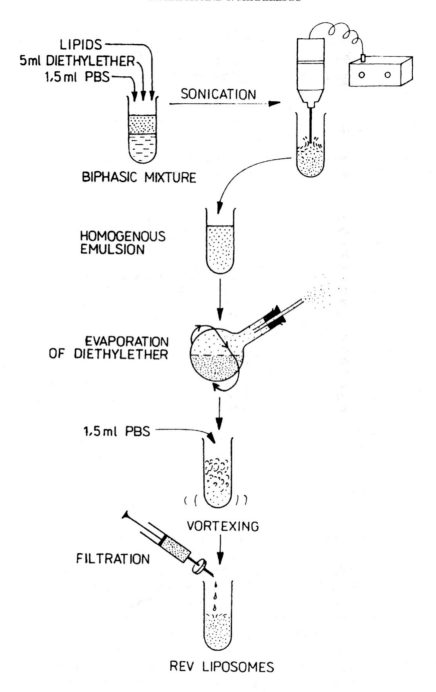

Figure 1 Schematic diagram of the system for preparation of reverse-phase vesicles (REV liposomes).

additional vortexing of the emulsion for 5 min, diethylether was evaporated *in vacuo*, at 20 °C, using a rotary evaporator. PBS (1.5 ml) was added to the resulting gel, which was stirred with a vortex mixer for another 10 min. The final suspension of liposomes was extruded through a Millipore membrane. To remove free, nonencapsulated peroxidase, 2 ml of liposomes suspension were passed through a Sephadex G-200 column (1.5 cm × 30 cm) eluted with PBS at pH 7.4. With this modified method, enzymatic activity was completely preserved, as shown in our previous experiments with more labile enzymes such as restriction endonucleases (Mixich, 1991).

The liposomes were administered intravascularly to white rats, both *in situ* and in isolated perfused hearts.

Animal Experiment Procedures

In situ experiments

Adult white rats (b.w. = 250–300 g) were anaesthetized with chloral hydrate (30 mg/100 g b.w.) and securely positioned on a surgical table. A midline incision of the abdominal skin was performed and the aorta and inferior vena cava were dissected. A thin polyethylene catheter was introduced inside the descending aorta, immediately above its bifurcation, and fixed in place by a fine suture. The inferior vena cava was cut and left open. Warm (37 °C) PBS was injected into the aorta, through the catheter, at a constant flow rate of 3 ml/min, by means of a peristaltic pump. After 5 min of perfusion, when the vasculature was washed free of blood, empty liposomes, (2 ml diluted with 7 ml of PBS) were introduced by perfusion and maintained in circulation for 3 min. The excess of liposomes was removed by perfusion with PBS for another 3 min. At the end of this interval, prefixation for electron microscopy was carried *in situ*, by perfusion for 3 min at a flow rate of 3 ml/min with 2% glutaraldehyde in 0.1 M sodium cacodylate buffer, pH 7.2.

At the end of the fixative perfusion, fragments of thoracic aorta were excised.

Isolated heart experiments

Adult white rats were anaesthetized intraperitoneally with chloral hydrate (30 mg/100 g b.w.). The trachea was cannulated and the thorax was opened under artificial ventilation. The ascending aorta was cannulated with a polyethylene cannula and perfused at a constant pressure of 75 mmHg, through an extension tube, according to the Langendorff technique. In order to avoid overloading the right heart with perfusate, the pulmonary artery was cut and left open. The isolated heart was excised from the thorax and connected directly to the perfusion system. We used as perfusate the Krebs-Henseleit buffer with the following composition: NaCl (127.2 mM), KCl (4.7 mM), $CaCl_2$ (2.5 mM), KH_2PO_4 (1.2 mM), $NaHCO_3$ (24.9 mM) and glucose (5.5 mM). The perfusate was oxygenated and maintained at a constant pH of 7.4 by bubbling with a gas mixture formed of 95% O_2/5% CO_2. The temperature of the perfusate was maintained at 38 °C by means of a heat exchanger and was continually measured at the level of the aortic cannula. The partial pressure of oxygen (pO_2), and the pH of the perfusate were periodically (every 5 min) measured with a pH and blood gas analyser (Instrumentation Laboratory IL-813) respectively. The value of the pO_2 was 500–550 mmHg.

After an equilibration time of 15–20 min, 0.5 ml suspension of liposomes with encapsulated peroxidase was injected in the perfusion fluid, through the aortic cannula, at a rate of 0.2 ml/minute. The heart was afterwards perfused for another 3 min with Krebs-Henseleit buffer only and specimens of myocardium were excised for electron microscopy processing.

Electron microscopy

The collected specimens of aorta obtained from *in situ* experiments were further fixed for 90 min by immersion in 2% glutaraldehyde in 0.1 M sodium cacodylate buffer, pH 7.2, post-fixed for 90 min at 4 °C in 1% OsO_4 in 0.1 sodium cacodylate (cacodylic acid, sodium salt) buffer pH 7.2, then treated *en bloc* with 0.5% uranyl acetate for 30 min at 20 °C, and then dehydrated and Epon™-embedded by standard procedures.

Preparation of specimens from isolated hearts was carried out generally by the same procedure described above, but with two major modifications: (a) the step of fixative perfusion was omitted, (b) after post-fixation with osmium tetroxide and prior to *en bloc* contrasting with uranyl acetate, a step for ultracytoenzymatic determination of peroxidase was introduced. Thus, heart specimens were incubated after extensive post-fixation washing in an incubation medium composed of: Tris-HCl buffer (0.05 M), pH 7.6 (10 ml); 3.3'-diaminobenzidine (5.0 mg) and 1% H_2O_2 (0.1 ml), for 30 min. After cytoenzymatic reaction, specimens were washed with Tris-HCl buffer, and processed as described above.

Ultrathin sections of tissue were cut with a Tesla (Czechoslovakia) ultramicrotome, and placed on 200-mesh copper grids, examined and micrographed under a Tesla BS-500 (Czechoslovakia) electron microscope.

RESULTS

The aorta specimens, excised after 3 min of contact with liposomes, were processed for electron microscopy in order to show the interaction of liposomes with the luminal side of the continuous endothelium. We did not consider it necessary to use marked liposomes because the vasculature was washed free of blood and therefore the liposomes can be distinguished from membranous structures of cellular origin was avoided. The liposomes of 300 nM diameter appear as oligolamellar vesicles. In Figure 2, all stages of endocytosis of liposomes by the endothelium can be observed. The fate of liposomes inside the endothelial cells after endocytosis could not be determined because the liposomes were not marked.

Liposomes marked with peroxidase were used in isolated heart experiments. After 3 min of contact with the coronary endothelium, these liposomes could be found in the extravascular spaces, both in the interstitial space (Figure 3a) and inside myocardial cells (Figures 3b–d). The passage of liposomes through the coronary capillary wall was a rapid process, because the interaction of the vesicles with the endothelium could not be shown after 3 minutes.

DISCUSSION AND CONCLUDING REMARKS

The results reported in this paper show that the access of liposomes to myocardial cells, after intravascular perfusion in isolated hearts, is a rapid process: after 3 min liposomes could be found in extravascular spaces. Multiple sections from atrial and ventricular muscle were examined. The results were identical in all experiments, i.e. liposomes were found both in the extracellular space and in cytoplasmic vacuoles in association with subcellular structures.

In agreement with our results, other investigators reported that liposomes were taken up by the dog heart *in vivo*, rabbit heart *in vitro*, and by cultured rat cardiomyocytes (Caride and Zaret, 1977; Gross and Sobel, 1979; Mueller *et al.*, 1980, 1981). These data indicate that intravascularly injected liposomes, containing a great variety of substances, can cross blood vessels with continuous endothelium, to reach cells of extravascular tissues.

Several possible mechanisms may be postulated to explain this process (Kayawake and Kako, 1982). Although many authors maintain that plasma membrane of endothelial cells is unable to 'flow' around a liposome (or another particle) bound to the cell surface, in the way that actively phagocytic cells engulf particles by surrounding them with pseudopodial extensions (Donald, 1980), we show by *in situ* experiments that endothelial cells from aorta can phagocytose liposomes with diameters of 300 nM. The existence of this mechanism was also morphologically demonstrated on confluent monolayers of cultured calf aortic endothelial cells (Zipper *et al.*, 1988).

Thus, the present results support the hypothesis that substances encapsulated into liposomes, are able to be transferred to tissues through a continuous endothelium barrier.

The major question remains whether a drug transported by liposomes can cross the blood-brain barrier by the above described mechanism. A large number of data from literature suggest that this is possible (Naoi and Yagi, 1980; Nagai, 1982; Yagi *et al.*, 1982; Osanai and Nagai, 1984; Kito *et al.*, 1987, 1989). These data, together with the results obtained in our laboratory using the experimental protocol described in this chapter, support the view that liposomes can pass through the BBB to reach brain cells.

The microvasculature of the myocardium is structurally representative for most of the microvascular beds of the organism with continuous (nonfenestrated) endothelium. Continuous endothelia have in common a large population of plasmalemmal

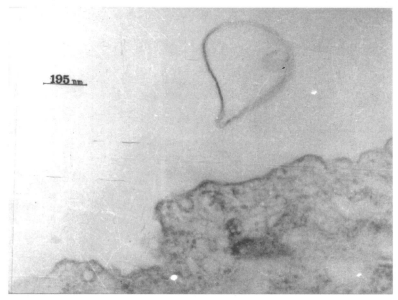

(2A)

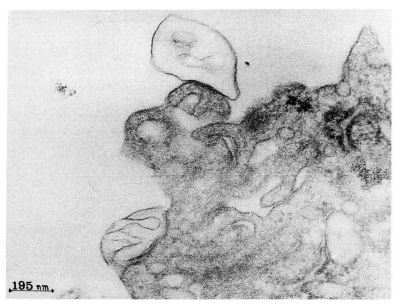

(2B)

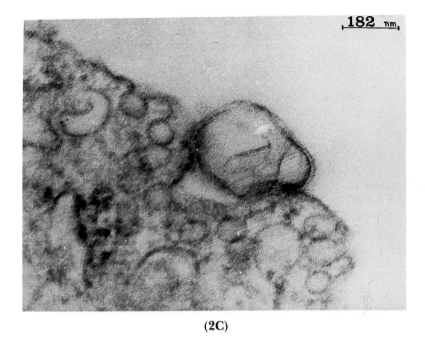

(2C)

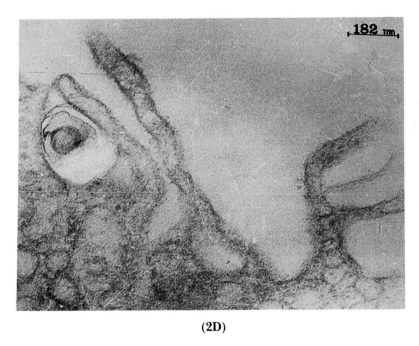

(2D)

Figure 2 (a–d). Four stages of phagocytosis of liposomes by the aorta endothelium cells.

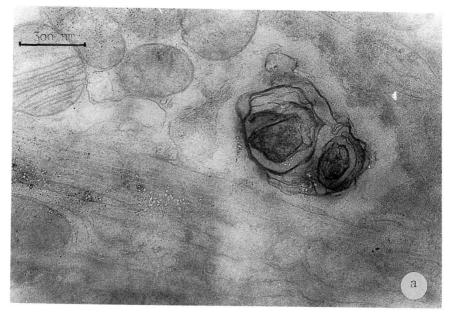

(3A)

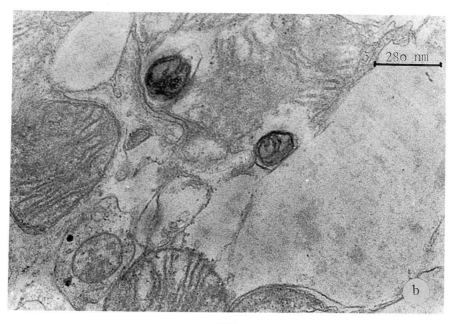

(3B)

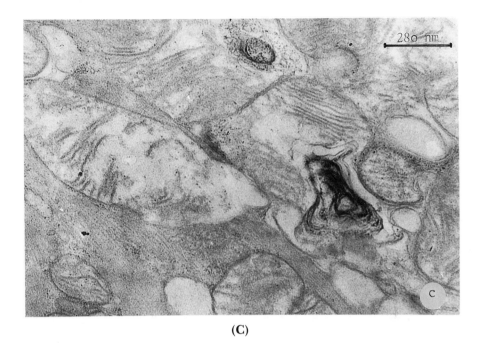

(C)

Figure 3 Peroxidase-positive liposomes found in the extravascular space (a), and inside myocardial cells (b, c) of isolated hearts.

vesicles (Palade *et al.*, 1979; Wagner and Casley-Smith, 1981) and intercellular tight junctions (Palade *et al.*, 1979; Simionescu, 1983). Vascular beds of this type have been used extensively in the past for physiological studies on capillary permeability. As a result, current concepts, theories, and hypotheses that pertain to this important topic rely primarily on data obtained on this type of microvasculature (Landis and Pappenheimer, 1963; Pietra *et al.*, 1982).

REFERENCES

Allen, T.M. (1994) Long-circulating (sterically stabilized) liposomes for targeted drug delivery. *Trends Pharmacol. Sci.*, **15**(7), 215–220.

Bangham, A.D., Hill, M.W. and Miller, N.G.A. (1974) Preparation and use of liposomes as models of biological membrane. In Korn, E.D. (Ed.), *Methods in membrane biology*, Vol. **1**, Plenum Press, New York, pp. 1–68.

Bankert, R.B., Mayhew, E., Yokota, S. and Jou, Y.H. (1985) Immunospecific tumor-targeting of drug-containing liposomes. In Milgrom, Abeyounis and Albini (Eds.), *Antibodies: Protective, destructive and regulatory role*, Karger, Basel, pp. 416–423.

Caride, V.J. and Zaret, B.L. (1977) Liposome accumulation in regions of experimental myocardial infarction. *Science*, **198**, 735–738.

Donald, K. (1980) Ultrastructure of reticuloendothelial clearance. In Carr, I. and Daems, W.T. (Eds.), *The reticuloendothelial system: A comprehensive treatise*, Plenum Press, New York, pp. 525–554.

Fendler, J.H. (1980) Optimizing drug entrapment in liposomes. Chemical and biophysical considerations. In Gregoriadis, G. and Allison, A.C. (Eds.), *Liposomes in biological systems*. J. Wiley & Sons, Chichester, pp. 87–100.

Foldvari, M., Faulkner, G.T. and Mezei, M. (1988) Imaging liposomes at electron microscopic level: Encapsulated colloidal iron as an electrondense marker for liposome-cell interactions. *J. Microencapsulation*, **5**(3), 231–241.

Gregoriadis, G., Kirby, C. and Senior, J. (1983) Optimization of liposome behavior *in vivo*. *Biol. Cell*, **47**(1), 11–18.

Gregoriadis, G.E. (Ed.) (1984) *Liposome technology*, Vol. 1, CRC Press, Boca Raton, Fla.

Gross, R.W. and Sobel, B.E. (1979) Augmentation of cardiac phospholipase activity induced with negative liposomes. *Trans. Assoc. Am. Phys.*, **92**, 136–147.

Hayakawa, T., Ushio, Y., Mogami, H. and Horibata, K. (1974a) The uptake, distribution and anti-tumor activity of bleomycin in gliomas in the mouse. *Eur. J. Cancer*, **10**, 137–142.

Hayakawa, T., Ushio, Y., Morimoto, K., Hasegawa, H., Mogami, H. and Horibata, K. (1974b) Uptake of bleomycin by human brain tumors. *J. Neurol. Neurosurg. Psychiatry*, **39**, 341–349.

Holmberg, E.G. and Huang, L. (1989) pH-sensitive and target sensitive immunoliposomes for drug targeting. In *Liposomes in the therapy of infectious diseases and cancer*, Alan R. Liss, Inc., pp. 25–34.

Kayawake, S. and Kako, K.J. (1982) Association of liposomes with the isolated perfused rabbit heart. *Basic Res. Cardiol.*, **77**, 668–681.

Kito, A., Yoshida, J., Kagyama, N., Kojima, N. and Yagi, K. (1987) Basic studies on chemotherapy of brain tumors by means of liposomes: Affinity of sulfatide-inserted liposomes to human glioma cells. *No To Shinkei*, **39**, 783–788.

Kito, A., Yoshida, J., Kageyama, N., Kojima, N. and Yagi, K. (1989) Liposomes coupled with monoclonal antibodies against glioma-associated antigen for targeting chemotherapy of glioma. *J. Neurosurg.*, **71**, 382–387.

Landis, E.M. and Pappenheimer, J.R. (1963) Exchange of substances through the capillary walls. In Hamilton, W.F. and Dow, P. (Eds.), *Handbook of Physiology, Sect. 2, Circulation II*, Amer. Physiol. Soc., Washington DC, pp. 961–1034.

Leserman, L.D., Marchy, P. and Barbet, J. (1984) Covalent coupling of monoclonal antibodies and protein A to liposomes: Specific interaction with cells *in vitro* and *in vivo*. In Gregoriadis, G. (Ed.), *Liposome technology*, Vol. III, CRC Press, Boca Raton, pp. 29–40.

Mixich, F. (1991) Induction of chromosomal aberrations by restriction endonucleases encapsulated in liposomes. *Mutation Res.*, **262**, 177–181.

Mueller, T., Hernsmeyer, K., Mayer, H. and Marcus, M. (1980) Conditions for liposome uptake by myocardial cells. *Amer. J. Cardiol.*, **45**, 414.

Mueller, T.M., Marcus, M.L., Mayer, H.E., Williams, J.K. and Hernsmeyer, K. (1981) Liposome concentration in canine ischemic myocardium and depolarized myocardial cells. *Circulation Res.*, **49**(2), 405–415.

Nagai, Y. (1982) Suppression of demyelination in acute EAE: New strategies for therapy of EAE and MS. In Kuroiwa, Y. and Kurland, L.T. (Eds.), *Multiple sclerosis east and west*, Kyushu Univ. Press, Fukuoka, Japan, pp. 347–358.

Naoi, M. and Yagi, K. (1980) Incorporation of enzyme through blood-brain barrier into the brain by means of liposomes. *Biochem. Intern.*, **1**(6), 591–596.

Nicolau, C. and Cudd, A. (1989) Liposomes as carriers of DNA. *Critical Rev. Therap. Drug Carrier Syst.*, **6**(3), 239–271.

Osanai, T. and Nagai, Y. (1984) Suppression of experimental allergic encephalomyelitis (EAE) with liposome-encapsulated protease inhibitor: Therapy through the blood-brain barrier. *Neurochem. Res.*, **9**(10), 1407–1415.

Pagano, R.E. and Weinstein, J.N. (1978) Interaction of liposomes with mammalian cells. *Annu. Rev. Biophys. Bioeng.*, **7**, 435–468.

Palade. G., Simionescu, M. and Simionescu, N. (1979) Structural aspects of the permeability of the microvascular endothelium. *Acta Physiol. Scand.*, **463**, 11–32.

Pietra, G.G., Fishman, A.P., Lanken, P.N., Sampson, P. and Hansen-Flaschen, J. (1982) Permeability of pulmonary endothelium to neutral and charged macromolecules. *Ann. Acad. Sci.*, **401**, 241–247.

Poste, G., Papahadjopoulos, D. and Vail, W.J. (1976) Lipid vesicles as carriers for introducing biologically active materials into cells. In: Prescott, D.M. (Ed.), *Methods in cell biology*, **14**, Academic Press, New York, pp. 33–71.

Poste, G. (1983) Liposome targeting *in vivo*. *Biol. Cell*, **47**(1), 19–38.

Riaz, M., Weiner, N. and Martin, F. (1988) Liposomes. In Lieberman, H.A., Rieger, M.M. and Banker, G.S. (Eds.), *Pharmaceutical dosage forms*, **2**, Marcel Decker, Inc., New York, Basel, pp. 567–603.

Ryman, B.F. and Tyrell, D.A. (1979) Liposomes-methodology and applications. In Dingle, J.T., Jacques, P.J. and Shaw, I.H. (Eds.), *Lysosomes in applied biology and therapeutics*, **6**, North Holland, Amsterdam, pp. 549–574.

Simionescu, N. (1983) Cellular aspects of transcapillary exchange. *Physiol. Rev.*, **63**, 1538–1579.

Storm, G., Steerenberg, P.A., van Borssum Waalkes, M., Emmen, F. and Crommelin, D.J.A. (1988) Potential pitfalls in *in vitro* antitumor activity testing of free and liposome-entrapped doxorubicin. *J. Pharm. Sci.*, **77**(10), 823–830.

Szoka, F.Jr. and Papahadjopoulos, D. (1978) Procedure for preparation of liposomes with large internal aqueous space and high capture by reverse-phase evaporation. *Proc. Natl. Acad. Sci. USA*, **75**, 4194–4198.

Takada, G., Onodera, H. and Tada, K. (1981) Tissue distribution of unentrapped or liposome-entrapped ^{131}I-labeled β-galactosidase injected into rats. *Tohoku J. Exp. Med.*, **134**, 103–114.

Toffano, G. and Bruni, A. (1980) Pharmacological properties of phospholipid liposomes. *Pharm. Res. Commun.*, **12**(9), 829–845.

Wagner, R.C. and Casley-Smith, J.R. (1981) Endothelial vesicles. *Microvasc. Res.*, **21**, 267–298.

Waser, P.G., Muller, U., Kreuter, J., Berger, S., Munz, K., Kaiser, E. and Pfluger, B. (1987) Localization of colloidal particles (liposomes, hexylcyanoacrylate nanoparticles and albumin nanoparticles) by histology and autoradiography in mice. *Intern. J. Pharm.*, **39**, 213–227.

Weissman, G. and Finkelstein, M. (1980) Uptake of enzyme-bearing liposomes by cells *in vivo* and *in vitro*. In Gregoriadis, G. and Allison, A.C. (Eds.), *Liposomes in biological systems*, J. Wiley & Sons, Chichester, pp. 153–178.

Yagi, K., Naoi, M., Sakai, H., Abe, H., Konishi, H. and Arichi, S. (1982) Incorporation of enzyme into the brain by means of liposomes of novel composition. *J. Appl. Biochem.*, **4**, 121–125.

Zipper, J., Sarrach, D. and Halle, W. (1988) Interaction of liposomes with vascular endothelial cells. *Biomed. Biochim. Acta*, **47**(7), 713–719.

PERSPECTIVES

A.G. de BOER AND W. SUTANTO

The various contributions to this book show several research interests which have the *in vitro* or *in vivo* blood-brain barrier (BBB) in common. The *in vitro* procedures comprise the culture of microvascular endothelial cells from brains of cow, calf, pig and rat while the *in vivo* approaches comprise microdialysis and voltammetry.

The *in vitro* procedures can be grossly divided into two well validated techniques to isolate brain microvessel endothelial cells (BMEC). Systems as published by Méresse *et al.* (1989), Abbott *et al.* (see Part A, II.1) and Rubin *et al.* (1991) may be considered in this way and have been proven to be reliable and powerful. These systems exist of co-cultures with astrocytes and use astrocyte conditioned medium respectively, and have been well characterised and are presently used by several participants of the Concerted Action: Drug Transport to the Brain: New experimental techniques. Nevertheless, these systems still do not fully represent the BBB since several other cell types (e.g. pericytes, neurones), the basement membrane and hormones from the blood may influence its functionality. This could lead to the conclusion that a complete artificial BBB system should be developed. However, this would be a very complicated system and therefore unpractical. The most pragmatic solution to this problem might be to look for that particular BBB-model that permits one to study a certain particular aspect of drug transport across the BBB. Once these *in vitro* studies have been performed, *in vivo* studies should be carried out as soon as possible.

Microdialysis and voltammetry methods may be applied for the *in vivo* studies. Microdialysis seems to be the method of choice for the quantitative measurement of drug transport across the BBB. *In vivo* recovery seems to be one of the problems, however various (robust) methods have been published to solve this problem (Lennroth, Jansen and Smith, 1987; Morrison *et al.*, 1991; Olson and Justice, 1993). Additional problems (see also chapter E.C.M. de Lange *et al.* in this book) comprise the recovery time following surgery before performing a microdialysis experiment and the influence of the introduction of the microdialysis probe on (brain) tissue and its effect on the microdialysis results. Other problems may arise with lipophilic drugs (adsorption), sensitivity of the detection method (preferentially electro-chemical and fluorescence detection) when very low concentrations of drugs or compounds have to be measured. At the moment most of these problems may be sufficiently solved. Voltammetry is a very interesting procedure to measure compounds that can be oxidized or reduced. A drawback may be the difficulty to reliably quantitate the measured compounds.

In general one may say that quite sophisticated *in vitro* and *in vivo* procedures are available to study BBB transport. Based on the wishes and demands of the particular researcher one may choose for a certain *in vitro* technique. The most important issue is that the technique is well validated. Within this concerted action several contributions have been made to achieve this, nevertheless this should be carefully checked in every new experimental setup. This may allow us to make predictions based on *in vitro* experimental results, however these should be verified as soon as possible *in vivo*.

REFERENCES

Méresse, S., Dehouck, M.-P., Delorme, P., Bensaïd, M., Tauber, J.-P., Delbart, C., Fruchart, J.-C. and Cecchelli, R. (1989) Bovine brain endothelial cells express tight junctions and monoamine oxidase activity in long-term culture. *J. Neurochem.*, **53**, 1363–1371.

Rubin, L.L., Hall, D.E., Parter, S., Barbu, K., Cannon, C., Horner, H.C., Janatpour, M., Liaw, C.W., Manning, K., Morales, J., Tanner, L.I., Tomaselli, K.J. and Bard, F. (1991) A cell culture model of the Blood-Brain Barrier. *The Journal of Cell Biology*, **115**, 1725–1735.

Lennroth, P., Jansen, P.A. and Smith, U. (1987) A microdialysis method allowing characterization of intercellular water space in humans. *Am. J. Physiol.*, **253**, E228–E231.

Morrison, P.F., Bungay, P.M., Hsiao, J.K., Mefford, I.N., Dijkstra, K.H. and Dedrick, R.L. (1991) Quantitative microdialysis. In: Robinson, T.E. and Justice, J.B. (Eds.), *Microdialysis in Neuroscience*, pp. 47–79.

Olson, R.J. and Justice, J.B. Jr. (1993) Quantitative microdialysis under transient conditions. *Anal. Chem.*, **65**, 1017–1022.

INDEX

ACE   18
acetaminophen   157–162
AHF   104–7
alkaline phosphatase   38, 41, 49, 91
ALP   91, 97
angiotensin converting enzyme   18, 75
artemisinin   119, 121–124, 126–128, 130,
   132–3
astrocyte   15, 17, 19, 23–4, 37–8, 61, 215
astrocyte conditioned medium   23–4, 61,
   215
atenolol   157–162

Bandeiraea simplicifolia isolectin I-B4   49,
   85
biogenic amine metabolites   43, 165, 169,
   172
blood-brain barrier   3, 5–6, 15, 17, 27–8,
   33–4, 37–8, 43–4, 49, 59, 69, 76, 81, 88,
   91, 117, 137, 150, 189, 198, 201–2, 207,
   215
bovine brain microvessel endothelial cells
   17, 59, 81
brain microvessels   17–8, 28, 37, 45, 59,
   76, 81

cAMP   15, 19, 60–1, 63–4, 67
catecholamines   165, 167, 169, 192
caudate nucleus   173, 180
CD31   100, 102, 105, 107
cell line   20, 27, 30–1, 33–4, 52, 81, 83
characterization   37, 41, 43, 111
co-culture   3, 5–6, 15, 18–9, 88
complement killing   14
continuous endothelium   203, 206–7
corneal endothelial cells   28, 30
culture media   3, 44
culture of astrocytes   17, 23
CYP1A   38, 41, 44
CYP2B   38, 41
CYP2D   38

CYP2E1   38
cytochrome P450   38, 42, 44

difference method   124, 143–4
differentiation   27, 33–4, 91, 97
DiI-Ac-LDL   110–112
dispase   7, 9, 28, 30, 39, 49–51, 81–2, 85–6,
   91, 93–4
dorsal hippocampus   165–6, 169, 172
drug metabolizing enzyme activities   38, 44

electrical resistance   18, 34, 49, 76, 85,
   91, 96
electronmicroscopy   49
endothelial cell culture   3, 5, 59, 109,
   110, 112
endothelial markers   18, 97
enzymatic   5, 18–9, 27–8, 37, 39, 41–45,
   49, 52, 69, 75, 81, 85, 87, 91, 109, 112,
   189, 195, 197, 205
enzyme activity   33
extracellular matrix   9, 28, 30, 34, 69–71,
   76, 105–6, 112

factor VIII   19, 22, 28, 31, 41, 49, 75, 85,
   97
fibroblast   53, 70, 88, 103–105
fibronectin   19, 28, 60, 64, 66, 100–1,
   104, 107, 112
Ficoll   11, 85

GnRH   179–183

HBEC   101, 106–7
histology   167, 180, 183
HSVEC   101, 106–7
human brain microvessel endothelium
   109–10

HUVEC   101, 106–7
hypotonic   157–9, 161–2

immortalized   20, 27–8, 30, 34, 52, 81, 83
immunoadsorption   101–2, 104, 106–7
immunohistochemical   49, 53, 88, 106
in vitro recovery   1, 3, 5, 15, 17–20, 27,
    34, 37–8, 43, 53, 59, 69–70, 76, 81, 88,
    91, 97, 99–100, 107, 119–20, 122–4,
    126–8, 132–3, 144, 151–2, 154, 161,
    166, 168, 189, 192, 195, 197, 202, 207,
    215–6
in vivo recovery   132, 154, 215
indolamines   165, 167, 169
isolation   3, 14, 17–21, 23, 27–8, 30, 34,
    49, 53, 59–60, 64, 66, 69, 81, 85, 93,
    99–100, 109–10
isolation procedure   21, 53, 64, 69, 85
isotonic   8, 12–3, 157–9, 161–2, 180

limitations   17, 59, 76, 119, 133
liposomes   201–3, 205–7
low perfusion rate   120, 126

Mab 5.6E   102, 104–7
MAO   18–9, 45
marker   15, 28, 31, 33, 100, 107, 110, 189,
    202–3
mass spectrometer   140
materials   3, 7, 17, 20, 22–3, 28, 38, 50,
    59–60, 70, 81, 85, 91–2, 100, 110, 121,
    137, 149, 158, 166, 175, 191, 203
mechanical   5, 18–9, 30, 69, 71, 75, 109–12,
    202
media   3, 31, 44, 52, 60, 70, 87, 92, 96,
    100, 103–5, 123, 154
metabolic barrier   37
monoclonal antibody   102
morphine   38, 119, 121–4, 126–8, 130,
    149, 151–5, 175
myocardium   201, 206–7

NADPH   38, 42, 44
neurotransmitter   45, 168
no-net flux   119–20, 124, 137, 144
nucleus accumbens   165, 167, 169, 172

P-glycoprotein   5, 14, 27–8, 30, 34
passaging   67, 103, 105
pericyte   14, 22–3, 75–6, 91, 97
perspectives   215
Pgp   27, 30
phagocytosis   201
pharmacodynamic   119–20
pharmacokinetic   119–20, 145, 152, 154,
    165, 172
physical barrier   37, 43, 99, 106, 201
PK/PD   119, 121, 130, 132–3
plating   6, 13, 25, 49, 51, 65–6, 71, 87,
    92, 94, 104
polyamine   100, 103
possibilities   17, 59
post-surgery interval   158
preparation   5–6, 9–10, 13, 15, 28, 44, 60,
    69–71, 81–2, 85, 91–4, 99, 102, 111,
    122, 128, 142, 203, 206
primary culture   6, 28, 30, 51, 76, 81, 85,
    87, 91, 109, 111–2
probe   112, 119–24, 126–30, 132–3, 138–9,
    141–4, 149–52, 154, 157–9, 161–2,
    166–9, 172–3, 177, 179–81, 183, 215
problem sources   14, 44, 52, 75, 88, 132
prostacyclin   102, 107, 109
prostaglandin   102
protein binding   119–20, 128

quality control   14, 44, 52, 75, 88, 97, 132

rat brain microvessel endothelial cells   17,
    20, 27
reagents   20, 42, 59–60, 101
reference method   143, 146
relative recovery   76, 119–20, 122–4, 126–8,
    130, 154, 168, 191
retinoic acid   27, 33, 34
retrodialysis   119–20, 123, 133, 137, 143
RIA   49, 85, 137, 140, 149, 151–2, 154

serum batch   103
smallest unilamillar veiscles   165, 168,
    172, 201
solutions   7, 60–1, 63, 66, 92, 121–2, 127,
    138, 166–7

spermine   100, 103, 105
sulfotransferase   44
surgery   128, 140, 142, 150, 157–8, 160–2,
   166–7, 175, 177, 183, 215

TEER   18–9, 49, 52, 56, 67, 76, 85, 88,
   96–7
temperature   25, 39, 52, 64, 66, 94–5,
   102, 110–1, 122, 129, 140, 157–9, 161–2,
   175, 203, 205

tonicity   157, 161
transcerebral microdialysis   149
trypsinization   22, 24, 31, 66–7

UEA–1   102, 107
uric acid   189, 192, 198

von Willebrand Factor   18, 28, 31, 100,
   109–12